Lecture Notes in Economics and Mathematical Systems

(Vol. 1–15: Lecture Notes in Operations Research and Mathematical Economics, Vol. 16–59: Lecture Notes in Operations Research and Mathematical Systems) For information about Vols. 1–29, please contact your bookseller or Springer-Verlag

Vol. 30: H. Noltemeier, Sensitivitätsanalyse bei diskreten linearen Optimierungsproblemen. VI, 102 Seiten. 1970.

Vol. 31: M. Kühlmeyer, Die nichtzentrale t-Verteilung. II, 106 Seiten. 1970.

Vol. 32: F. Bartholomes und G. Hotz, Homomorphismen und Reduktionen linearer Sprachen. XII, 143 Seiten. 1970. DM 18,–

Vol. 33: K. Hinderer, Foundations of Non-stationary Dynamic Programming with Discrete Time Parameter. VI, 160 pages. 1970.

Vol. 34: H. Störmer, Semi-Markoff-Prozesse mit endlich vielen Zuständen. Theorie und Anwendungen. VII, 128 Seiten. 1970.

Vol. 35: F. Ferschl, Markovketten. VI, 168 Seiten. 1970.

Vol. 36: M. J. P. Magill, On a General Economic Theory of Motion. VI, 95 pages. 1970.

Vol. 37: H. Müller-Merbach, On Round-Off Errors in Linear Programming. V, 48 pages. 1970.

Vol. 38: Statistische Methoden I. Herausgegeben von E. Walter. VIII, 338 Seiten. 1970.

Vol. 39: Statistische Methoden II. Herausgegeben von E. Walter. IV, 157 Seiten. 1970.

Vol. 40: H. Drygas, The Coordinate-Free Approach to Gauss-Markov Estimation. VIII, 113 pages. 1970.

Vol. 41: U. Ueing, Zwei Lösungsmethoden für nichtkonvexe Programmierungsprobleme. IV, 92 Seiten. 1971.

Vol. 42: A. V. Balakrishnan, Introduction to Optimization Theory in a Hilbert Space. IV, 153 pages. 1971.

Vol. 43: J. A. Morales, Bayesian Full Information Structural Analysis. VI, 154 pages. 1971.

Vol. 44: G. Feichtinger, Stochastische Modelle demographischer Prozesse. IX, 404 Seiten. 1971.

Vol. 45: K. Wendler, Hauptaustauschschritte (Principal Pivoting). II, 64 Seiten. 1971.

Vol. 46: C. Boucher, Leçons sur la théorie des automates mathématiques. VIII, 193 pages. 1971.

Vol. 47: H. A. Nour Eldin, Optimierung linearer Regelsysteme mit quadratischer Zielfunktion. VIII, 163 Seiten. 1971.

Vol. 48: M. Constam, FORTRAN für Anfänger. 2. Auflage. VI, 148 Seiten. 1973.

Vol. 49: Ch. Schneeweiß, Regelungstechnische stochastische Optimierungsverfahren. XI, 254 Seiten. 1971.

Vol. 50: Unternehmensforschung Heute – Übersichtsvorträge der Züricher Tagung von SVOR und DGU, September 1970. Herausgegeben von M. Beckmann. IV, 133 Seiten. 1971.

Vol. 51: Digitale Simulation. Herausgegeben von K. Bauknecht und W. Nef. IV, 207 Seiten. 1971.

Vol. 52: Invariant Imbedding. Proceedings 1970. Edited by R. E. Bellman and E. D. Denman. IV, 148 pages. 1971.

Vol. 53: J. Rosenmüller, Kooperative Spiele und Märkte. III, 152 Seiten. 1971.

Vol. 54: C. C. von Weizsäcker, Steady State Capital Theory. III, 102 pages. 1971.

Vol. 55: P. A. V. B. Swamy, Statistical Inference in Random Coefficient Regression Models. VIII, 209 pages. 1971.

Vol. 56: Mohamed A. El-Hodiri, Constrained Extrema. Introduction to the Differentiable Case with Economic Applications. III, 130 pages. 1971.

Vol. 57: E. Freund, Zeitvariable Mehrgrößensysteme. VIII,160 Seiten. 1971.

Vol. 58: P. B. Hagelschuer, Theorie der linearen Dekomposition. VII, 191 Seiten. 1971.

Vol. 59: J. A. Hanson, Growth in Open Economies. V, 128 pages. 1971.

Vol. 60: H. Hauptmann, Schätz- und Kontrolltheorie in stetigen dynamischen Wirtschaftsmodellen. V, 104 Seiten. 1971.

Vol. 61: K. H. F. Meyer, Wartesysteme mit variabler Bearbeitungsrate. VII, 314 Seiten. 1971.

Vol. 62: W. Krelle u. G. Gabisch unter Mitarbeit von J. Burgermeister, Wachstumstheorie. VII, 223 Seiten. 1972.

Vol. 63: J. Kohlas, Monte Carlo Simulation im Operations Research. VI, 162 Seiten. 1972.

Vol. 64: P. Gessner u. K. Spremann, Optimierung in Funktionenräumen. IV, 120 Seiten. 1972.

Vol. 65: W. Everling, Exercises in Computer Systems Analysis. VIII, 184 pages. 1972.

Vol. 66: F. Bauer, P. Garabedian and D. Korn, Supercritical Wing Sections. V, 211 pages. 1972.

Vol. 67: I. V. Girsanov, Lectures on Mathematical Theory of Extremum Problems. V, 136 pages. 1972.

Vol. 68: J. Loeckx, Computability and Decidability. An Introduction for Students of Computer Science. VI, 76 pages. 1972.

Vol. 69: S. Ashour, Sequencing Theory. V, 133 pages. 1972.

Vol. 70: J. P. Brown, The Economic Effects of Floods. Investigations of a Stochastic Model of Rational Investment. Behavior in the Face of Floods. V, 87 pages. 1972.

Vol. 71: R. Henn und O. Opitz, Konsum- und Produktionstheorie II. V, 134 Seiten. 1972.

Vol. 72: T. P. Bagchi and J. G. C. Templeton, Numerical Methods in Markov Chains and Bulk Queues. XI, 89 pages. 1972.

Vol. 73: H. Kiendl, Suboptimale Regler mit abschnittweise linearer Struktur. VI, 146 Seiten. 1972.

Vol. 74: F. Pokropp, Aggregation von Produktionsfunktionen. VI, 107 Seiten. 1972.

Vol. 75: GI-Gesellschaft für Informatik e.V. Bericht Nr. 3. 1. Fachtagung über Programmiersprachen · München, 9.–11. März 1971. Herausgegeben im Auftrag der Gesellschaft für Informatik von H. Langmaack und M. Paul. VII, 280 Seiten. 1972.

Vol. 76: G. Fandel, Optimale Entscheidung bei mehrfacher Zielsetzung. II, 121 Seiten. 1972.

Vol. 77: A. Auslender, Problèmes de Minimax via l'Analyse Convexe et les Inégalités Variationelles: Théorie et Algorithmes. VII, 132 pages. 1972.

Vol. 78: GI-Gesellschaft für Informatik e.V. 2. Jahrestagung, Karlsruhe, 2.–4. Oktober 1972. Herausgegeben im Auftrag der Gesellschaft für Informatik von P. Deussen. XI, 576 Seiten. 1973.

Vol. 79: A. Berman, Cones, Matrices and Mathematical Programming. V, 96 pages. 1973.

Vol. 80: International Seminar on Trends in Mathematical Modelling, Venice, 13–18 December 1971. Edited by N. Hawkes. VI, 288 pages. 1973.

Vol. 81: Advanced Course on Software Engineering. Edited by F. L. Bauer. XII, 545 pages. 1973.

Vol. 82: R. Saeks, Resolution Space, Operators and Systems. X, 267 pages. 1973.

continuation on page 159

Lecture Notes
in Economics and
Mathematical Systems

Managing Editors: M. Beckmann and H. P. Künzi

181

Hanif D. Sherali
C. M. Shetty

Optimization with
Disjunctive Constraints

Springer-Verlag
Berlin Heidelberg New York 1980

Authors

Hanif D. Sherali
School of Industrial Engineering
and Operations Research
Virginia Polytechnic Institute
and State University
Blacksburg, VA 24061/USA

C. M. Shetty
School of Industrial and Systems Engineering
Georgia Institute of Technology
Atlanta, GA 30332/USA

AMS Subject Classifications (1980): 05 A 20, 49 D 37, 90-02, 90 C 10, 90 C 11, 90 C 33

ISBN-13: 978-3-540-10228-1   e-ISBN-13: 978-3-642-48794-1
DOI: 10.1007/ 978-3-642-48794-1

Library of Congress Cataloging in Publication Data. Sherali, Hanif D 1952- Optimization with disjunctive constraints. (Lecture notes in economics and mathematical systems; 181) Bibliography: p. Includes index. 1. Mathematical optimization. I. Shetty, C. M., 1929- joint author. II. Title. III. Series. QA402.5.S53. 519.4. 80-20137

Printing and binding: Beltz Offsetdruck, Hemsbach/Bergstr.
2142/3140-543210

Dedicated to

Our Parents and Teachers

PREFACE

The disjunctive cut principle of Balas and Jeroslow, and the related polyhedral annexation principle of Glover, provide new insights into cutting plane theory. This has resulted in its ability to not only subsume many known valid cuts but also improve upon them. Originally a set of notes were written for the purpose of putting together in a common terminology and framework significant results of Glover and others using a geometric approach, referred to in the literature as convexity cuts, and the algebraic approach of Balas and Jeroslow known as Disjunctive cuts. As it turned out subsequently the polyhedral annexation approach of Glover is also closely connected with the basic disjunctive principle of Balas and Jeroslow. In this monograph we have included these results and have also added several published results which seem to be of strong interest to researchers in the area of developing strong cuts for disjunctive programs. In particular, several results due to Balas [4,5,6,7], Glover [18,19] and Jeroslow [23,25,26] have been used in this monograph. The appropriate theorems are given without proof. The notes also include several results yet to be published [32,34,35] obtained under a research contract with the National Science Foundation to investigate solution methods for disjunctive programs.

The monograph is self-contained and complete in the sense that it attempts to pool together existing results which the authors viewed as important to future research on optimization using the disjunctive cut approach. However, we have not attempted to record and discuss all important known valid inequalities, and methods to develop them. We have also listed only a minimum of references. An interested researcher will find readily a larger and more meaningful list of references in [4,5,6,7,18,19,20,23,25,26].

In writing this monograph and in reporting the research results, the authors found the works of Egon Balas, Fred Glover and Bob Jeroslow fundamental and extremely thought provoking. These publications initiated this study, and we are deeply indebted to them. We are also indebted to the National Science Foundation for supporting the research endeavor on Disjunctive Programming under their grant No. ENG 77-23683 and to Mike Thomas, Director of the School of Industrial Engineering at the Georgia Institute of Technology, for the support we have received in successfully completing this project. Finally, we are thankful to Mrs. Joene Owen for her cooperation and an excellent typing of this manuscript.

Hanif D. Sherali
School of Industrial Engineering
  and Operations Research
Virginia Polytechnic Institute and
  State University
Blacksburg, VA 24061

C. M. Shetty
School of Industrial and Systems
  Engineering
Georgia Institute of Technology
Atlanta, GA 30332

Table of Contents

Page

Chapter I: Introduction 1

1.1 Basic Concepts 1
1.2 Special Cases of Disjunctive Programs and Their
    Applications 4
1.3 Notes and References 10

Chapter II: Basic Concepts and Principles 12

2.1 Introduction 12
2.2 Surrogate Constraints 12
2.3 Pointwise-Supremal Cuts 15
2.4 Basic Disjunctive Cut Principle 16
2.5 Notes and References 18

Chapter III: Generation of Deep Cuts Using the Fundamental
             Disjunctive Inequality 19

3.1 Introduction 19
3.2 Defining Suitable Criteria for Evaluating the Depth of
    a Cut 22
3.3 Deriving Deep Cuts for DC1 24
3.4 Deriving Deep Cuts for DC2 31
3.5 Other Criteria for Obtaining Deep Cuts 51
3.6 Some Standard Choices of Surrogate Constraint
    Multipliers 53
3.7 Notes and References 54

Chapter IV: Effect of Disjunctive Statement Formulation on
            Depth of Cut and Polyhedral Annexation Techniques 55

4.1 Introduction 55
4.2 Illustration of the Tradeoff Between Effort for Cut
    Generation and the Depth of Cut 55
4.3 Some General Comments with Applications to the
    Generalized Lattice Point and the Linear Complementarity
    Problem 62
4.4 Sequential Polyhedral Annexation 64
4.5 A Supporting Hyperplane Scheme for Improving Edge
    Extensions 72
4.6 Illustrative Example 78
4.7 Notes and References 80

Chapter V: Generation of Facets of the Closure of the Convex Hull of
           Feasible Points 81

5.1 Introduction 81
5.2 A Linear Programming Equivalent of the Disjunctive
    Program 82
5.3 Alternative Characterization of the Closure of the Convex
    Hull of Feasible Points 85
5.4 Generation of Facets of the Closure of the Convex Hull
    of Feasible Points 92
5.5 Illustrative Example 104
5.6 Facial Disjunctive Programs 107
5.7 Notes and References 112

Page

Chapter VI: Derivation and Improvement of Some Existing Cuts
             Through Disjunctive Principles 113

    6.1 Introduction 113
    6.2 Gomory's Mixed Integer Cuts 113
    6.3 Convexity or Intersection Cuts with Positive Edge
         Extensions 118
    6.4 Reverse Outer Polar Cuts for Zero-One Programming 121
    6.5 Notes and References 126

Chapter VII: Finitely Convergent Algorithms for Facial Disjunctive
              Programs with Applications to the Linear Complementarity
              Problem 127

    7.1 Introduction 127
    7.2 Principal Aspects of Facial Disjunctive Programs 127
    7.3 Stepwise Approximation of the Convex Hull of Feasible
         Points 129
    7.4 Approximation of the Convex Hull of Feasible Points
         Through an Extreme Point Characterization 131
    7.5 Specializations of the Extreme Point Method for the
         Linear Complementarity Problem 136
    7.6 Notes and References 140

Chapter VIII: Some Specific Applications of Disjunctive Programming
               Problems 141

    8.1 Introduction 141
    8.2 Some Examples of Bi-Quasiconcave Problems 141
    8.3 Load Balancing Problem 146
    8.4 The Segregated Storage Problem 148
    8.5 Production Scheduling on N-Identical Machines 149
    8.6 Fixed Charge Problem 152
    8.7 Project Selection/Portfolio Allocation/Goal Programming . . 153
    8.8 Other Applications 153
    8.9 Notes and References 154

SELECTED REFERENCES 155

Chapter I

INTRODUCTION

1.1   Basic Concepts

A disjunctive program is an optimization problem where the constraints represent logical conditions. In this monograph we are concerned with such conditions expressed as linear constraints. The methods associated with disjunctive programming are by no means novel. Some of the methods proposed over two decades ago to solve integer programming problems used cutting planes derived from logical statements implying integrality. It can be shown that these problems can be viewed as disjunctive programs and the cutting planes used in integer programming are special applications of the principal theorem in disjunctive programming. As amply demonstrated by the recent works of Balas, Glover and Jeroslow, the disjunctive programming approach has provided a powerful unifying theory of all cutting plane solution strategies. Furthermore, it has provided a completely different perspective to examine this theory and has enabled one to derive deeper insights into existing knowledge. In the exposition that follows, we will be presenting the existing and new thoughts on disjunctive programming so that a reader can readily understand the developments thus far, and appreciate the potentials for research in this area.

Let us first introduce some fundamental concepts involved in our investigation. By the term disjunctive program, we mean a linear or nonlinear program which contains logical conditions stated as linear constraints. In our context, logical conditions include the following operations, stated in terms of say, conditions A and B.

(i) Conjunction - denoted by $A \wedge B$, this asserts that both conditions A and B must hold. As an example, a polyhedral set may be viewed as a conjunction of several linear inequalities or half spaces.

(ii) Disjunction - denoted by $A \vee B$, this asserts that either condition A or B (or both) must hold. A common example of this, as mentioned above,

arises in linear zero-one programs. There, in the presence of the
restriction $0 \leq x \leq 1$ on each variable, one has the disjunction that
either $x \leq 0$ or $x \geq 1$ must hold.

(iii) <u>Negation or Complement</u> – denoted by $\bar{A}$ this asserts that condition A
must not hold. For example, one might assert in some context that the
total cost $2x_1 + 3x_2$, say, must not exceed 7 units. Thus, condition
A is $2x_1 + 3x_2 \geq 7$ and the relevant negation is $2x_1 + 3x_2 \leq 7$.

(iv) <u>Implication</u> – denoted by A => B, this asserts that if A holds, then B
must hold. As an example one might say in some context that if a plant
i is located at a certain potential site, then the total output from it
must be at least p_i units. Letting $y_i = 1$ or 0 according as the plant
i is located or not, and letting $\sum_j x_{ij}$ denote the total output from it,
the implication condition is

$$y_i = 1 \Rightarrow \sum_j x_{ij} \geq p_i$$

Note that this implication is equivalent to the disjunction

$$\{y_i = 0\} \ V \ \{\sum_j x_{ij} \geq p_i\}$$

In general, A => B is equivalent to the disjunction $\bar{A} V B$.
Hence, examining the above logical conditions, one may note that conjunctions and
negations stated in terms of linear inequalities lead to polyhedral sets which
are, as it is well known, convex. Moreover, implications are essentially disjunctions as shown above. Now, it is the operation of disjunction which leads to
nonconvexities and renders the problem of interest to us.

Let us now proceed to formulate a disjunctive program in a general setting
and then cite and briefly examine several important problems which are special
cases of this problem. The notations we use throughout this study are, as far as
possible, consistent with those in existing literature.

Consider the following constraint sets S_h, where $h \in H$, an index set which
may or may not be of finite cardinality.

$$S_h = \{x: A^h x \geq b^h, \ x \geq 0\}, \ h \in H \tag{1.1}$$

In terms of the sets S_h, one may state a _disjunction_

$$x \in \bigcup_{h \in H} S_h \ \text{or simply} \ \bigvee_{h \in H} \{A^h x \geq b^h, \ x \geq 0\} \tag{1.2}$$

This disjunction may be imbedded into a general problem called a disjunctive pro-

gram as follows:

> DP: minimize $f(x)$
>
>   subject to $x \in X$
>
> $x \in \bigcup_{h \in H} \{A^h x \geq b^h, \ x \geq 0\}$

where $f: R^n \rightarrow R$ is lower semicontinuous and where X, is a closed subset of the

nonnegative orthant of R^n.

The application of _disjunctive methods_ to solve problems of type DP above,

involve the derivation of suitable cutting planes or valid linear inequalities

defined as follows:

Definition - An inequality $\pi x \geq \pi_0$ is said to be a _valid inequality_ for the dis-

junction $x \in \bigcup_{h \in H} S_h$ if

$$x \in S = \bigcup_{h \in H} S_h \ \text{implies} \ \pi x \geq \pi_0 \tag{1.3}$$

Before proceeding any further, let us pause and examine some special cases

of Problem DP which have been of interest to researchers. These problems, dis-

cussed along with their applications in the next section, include the generalized

lattice point problem, the cardinality constrained linear program, the binary

mixed-integer linear program, the extreme point optimization problem, the linear

complementarity problem, and numerous others. Later in Chapter VIII we will discuss

in some detail certain specific problems viewed as disjunctive programs.

1.2 Special Cases of Disjunctive Programs and Their Applications

1.2.1 The Generalized Lattice Point Problem

This problem may be stated mathematically as follows

GLPP: minimize $c^t x$

subject to $v = d - Dx \geq 0$ (1.4)

$u = b - Ax \geq 0$

$$\left\{ \begin{array}{l} \text{and at least q components of u, corresponding to linearly} \\ \text{independent rows of A, must be zero.} \end{array} \right\} \quad (1.5)$$

Note that the superscript t will, unless otherwise stated, be used to denote the matrix transpose operation. Here, we assume that D is of dimension mxn and A is of dimension pxn. Now let us take different combinations of q out of p components of u which correspond to linearly independent rows of A. Thus, suppose that there are $\hat{h} \leq \binom{p}{q}$ such combinations and let H denote the index set $\{1,\ldots,\hat{h}\}$. For any such combination, say $h \in H$, define the set

$S_h = \{u: u_i = 0$ if i is the index of one of the q components of

u corresponding to h$\}$, for $h \in H$ (1.6)

Then, Problem GLPP may be restated in a form, usually referred to as the disjunctive normal form, as follows:

minimize $c^t x$

subject to $v = d - Dx \geq 0$

$u = b - Ax \geq 0$

$u \in \bigcup_{h \in H} S_h$ (1.7)

Note that constraints (1.5) or equivalently, constraint (1.7) with S_h defined in (1.6), essentially states that u must be an interior point with respect to at most a p-q dimensional face of the set $U = \{u: u_i \geq 0$ for each i$\}$. We remark that one may relax constraint (1.5) to simply assert that at least q of the p components of u must be zero, whence, $\hat{h} = \begin{pmatrix} p \\ q \end{pmatrix}$.

Problem GLPP has been used as a special subroutine for minimizing a concave function over a convex region and for determining the most degenerate solution to a linear programming problem. In the latter context, such a solution is desirable, for example, in a fixed charge problem which has large fixed costs and linear variable costs. In this case, the most degenerate linear programming solution yields a good lower bound and/or starting point for any other scheme. Among other applications, the multiple choice problem is of significant importance.

1.2.2 The Cardinality Constrained Linear Program

This problem is a special case of the generalized lattice point problem, and may be stated as follows,

CCLP: minimize $c^t x$

subject to $Dx \leq d$

$x \geq 0$

$|x|^+ \leq n-q$ (1.8)

where D is of dimension mxn and $|x|^+$ denotes the number of positive components of the vector x. Again, as before, we may transform constraint (1.8) to restate the problem in the normal disjunctive form. For this purpose, define the index set $H = \{1,\ldots,\binom{n}{q}\}$ and let each $h \in H$ correspond to q particular components of x such that set H exhausts all such combinations. Hence, define

$S_h = \{x:$ the q components of x corresponding to h are equal to

zero$\}$ (1.9)

Thus, Equation (1.8) may be replaced by

$$x \in \bigcup_{h \in H} S_h$$ (1.10)

As an application, one may consider the manufacture of several (n) items at a production facility and let x_i denote the volume of production for item or product i, $i=1,\ldots,n$. The constraints $Dx \leq d$ may represent resource limitations and the disjunctive constraints (1.8), or equivalently (1.10) with (1.9), may restrict the production to at most (n-q) items.

In a like manner, one may be concerned with the location of (n-q) facilities at a subset of a number of potential sites. These facilities are required to satisfy a certain demand. The problem then may be to optimally locate these facilities and determine their capacity so as to minimize costs while satisfying demands.

There is a generalization of Problem CCLP known as the Element Constrained Linear Program (ECLP). Here, decision variables y_i, $i=1,\ldots,n$ are defined according to

$$y_i = \begin{cases} 1 & \text{if } x_i > 0 \\ 0 & \text{otherwise} \end{cases} \quad \text{for } i=1,\ldots,n$$

The decision vector $y=(y_1,\ldots,y_n)$ then is restricted according to a constraint set $Fy \leq f$. Thus, Problem CCLP is a special case of Problem ECLP with $Fy \leq f$ denoting the single constraint

$$\sum_{i=1}^{n} y_i \leq n-q$$

Hence, in the first example cited above for Problem CCLP, one may have certain contingency constraints between products or certain products may be mutually exclusive. Such interactions between products would convert the problem into an element constrained linear program.

1.2.3 The Binary Mixed Integer Linear Program

This problem is also a special case of Problem GLPP. It may be formulated mathematically as follows:

BMILP: minimize $c_1^t x_1 + c_2^t x_2$

               subject to $D_1 x_1 + D_2 x_2 \leq d$

                               $x_1 \geq 0$

                               $x_{2i} = 0$ or 1 for each $i=1,\ldots,n$ (1.11)

where D_2 is of dimension mxn. To write Problem BMILP in the disjunctive normal form, note that (1.11) is equivalent to the following constraints:

$$x_{2i} + u_i = 1 \qquad i=1,\ldots,n$$

$$x_{2i}, u_i \geq 0 \qquad i=1,\ldots,n$$

$$\{\text{At least n components of } (x_2,u) \text{ are zero}\} \qquad (1.12)$$

Now one may transform Equation (1.12) in a manner identical to that used for transforming (1.8) to (1.10) through the definition (1.9). Problem BMILP has several well known applications such as the multiple choice programming problem, the knapsack problem, the fixed-charge location-allocation problem, and others.

1.2.4 The Extreme Point Optimization Problem

This problem is closely related to Problem GLPP, and may be stated as follows:

EPP: minimize $c^t x$

subject to $Dx = d$

x is an extreme point of $P = \{x: Ax = b, x \geq 0\}$ (1.13)

Hence, D is of dimension mxn and A is of dimension (pxn). Let us attempt to re-write Equation (1.13). Consider any point $x \in X$ and identify those components x_j of this point x which satisfy $x_j = 0$. Let $J = \{j: x_j = 0\} \subseteq \{1,\ldots,n\}$. Now construct a matrix $\tilde{I}$ whose rows are comprised of unit vectors e_j, each row corresponding to a $j \in J$, where e_j has all components zero except for a unity in position j. Then consider the matrix $\binom{A}{\tilde{I}}$. Then one may easily see that a point $x \in X$ is an extreme point of X if and only if $\binom{A}{\tilde{I}}$ has rank n.

Thus, to write Problem EPP in a disjunctive normal form, consider an enumeration of all subsets J_h of the set $\{1,\ldots,n\}$ such that if one constructs a matrix I_h for each such J_h, where I_h has rows comprised of vectors e_j for $j \in J_h$, then the matrix $\binom{A}{I_h}$ has rank n. Further, let H contain the indices h corresponding to such sets J_h. Then let us define

$$S_h = \{x: \; x_j \le 0 \text{ for each } j \in J_h, \; x \ge 0\} \text{ for each } h \in H \qquad (1.14)$$

Using (1.14), we may now re-write (1.13) to formulate Problem EPP in a disjunctive normal form as follows:

$$\text{minimize} \quad c^t x$$
$$\text{subject to} \quad Dx = d$$
$$Ax = b$$
$$x \in \bigcup_{h \in H} S_h$$

Applications of Problem EPP include several bilinear programming problems such as the location-allocation problem using rectilinear distance measure. The problem of minimizing inventory and changeover costs for a single machine scheduling situation has also been formulated as Problem EPP. Another application is its use as a subroutine in a cutting plane procedure to find an extreme point of a set which is also feasible to a system of cuts generated at any stage.

1.2.5 The Linear Complementarity Problem

This problem may be stated mathematically as

$$\text{LCP:} \qquad \text{minimize} \quad c^t x$$
$$\text{subject to} \quad Dx = d$$
$$x \ge 0$$
$$x_p x_q = 0 \quad \text{for each } (p,q) \in Z$$

where Z is an appropriate set of two-tuple indices. Now, consider the construction of $2^{|Z|}$ distinct sets J_h, $h \in H = \{1,\ldots,2^{|Z|}\}$, where each J_h has exactly one of the indices p, q for each $(p,q) \in Z$. Define

$$S_h = \{x: \; x_j \le 0 \text{ for } j \in J_h, \; x \ge 0\} \quad \text{for each } h \in H.$$

Then Problem LCP may be restated in the disjunctive normal form as

$$\text{minimize} \quad c^t x$$

$$\text{subject to} \quad Dx = d$$

$$x \in \bigcup_{h \in H} S_h$$

When the cardinality of the set H is small, Problem DP can easily be solved using the solution of $|H|$ problem as shown by Theorem 1.1 below. When this direct approach is not available, we need more sophisticated tools. This is the subject of discussion over the next few chapters.

Theorem 1.1.

Consider Problem DP stated above and assume $|H| < \infty$. Define problems

$$DP_h: \quad \text{minimize} \{f(x): x \in X \cap S_h\} \quad \text{for each } h \in H \quad (1.15)$$

Let x^h solve DP_h. Then x^{h*} solves DP, where

$$f(x^{h*}) = \underset{h \in H}{\text{minimum}} \quad \{f(x^h)\} \quad (1.16)$$

Proof. By contradiction, suppose x^* solves DP with $f(x^*) < f(x^{h*})$, and assume that $x^* \in S_{\hat{h}}$ for some $\hat{h} \in H$. Since x^* is feasible to $DP_{\hat{h}}$ and $x^{\hat{h}}$ solves $DP_{\hat{h}}$, we must have $f(x^*) \geq f(x^{\hat{h}}) \geq f(x^{h*})$, a contradiction. This completes the proof.

Essentially, Theorem 1.1 involves the solution of $|H|$ problems in order to recover an optimal solution to Problem DP. This may be a viable approach for some special problems for which the cardinality of H is not too large. For example, one may be considering a production planning problem in which each set S_h may represent the restrictions on the process accruing from the implementation of production method $h \in H$. On the other hand, for zero-one linear integer programs for example, the application of Theorem 1.1 is tantamount to total enumeration and for a complementarity problem which requires, say, $u_j v_j = 0$ for $j=1,\ldots,m$, one would need to solve 2^m problems to obtain an optimal solution. It is for the solution of such problems, that we devote this study.

We now discuss some basic concepts and principles involved in disjunctive programming methods. An attempt is made in this chapter to present thoughts and ideas and to derive results so that the development is intuitively appealing. Thereafter, we discuss in a general context, the derivation of deep disjunctive cuts and also look at certain specializations. We then digress momentarily to demonstrate how the depth of cut that can be derived depends upon the formulation of the disjunctive statement. Based on this exposition, we discuss procedures for strengthening given valid cuts. This is then related to the supports and facets of the convex hull of feasible points. Following this, we show that disjunctive cutting planes subsume all other types of cutting planes by recovering several known cutting planes from a general form of a disjunctive cutting plane. Finally, we treat special cases of disjunctive programs. First, we demonstrate how the notion of the convex hull of feasible points admits two finitely convergent procedures for a special class of disjunctive programs known as facial disjunctive programs. Thereafter, we discuss, some specific applications.

1.3 Notes and References

Owen [29] considered a class of problems where at least one variable from each of several sets is required to be equal to zero. Applications of this formulation include integer programming, the linear complementary problem and the concave minimization problem. Since valid inequalities were derived from certain logical disjunctions, Owen called these valid inequalities disjunctive constraints. He noted that the cuts derived are indeed special cases of valid inequalities derived by Glover and Klingman [15] in the context of generalized Lattice Point problems. In his paper, Owen has really given a primitive algorithm, but the spirit of the approach is that of the cutting plane algorithms proposed by Gomory, Balas, Young and others for Integer Programming, by Tui, Balas, and Ritter in the context of nonconvex problems with linear constraints. It is covered by the general theory of convexity cuts of Glover [16,18], and overlaps with the work of Balas [2,3] and Burdet [10,12] using polarity.

An excellent survey of disjunctive programming principles and applications in the spirit of this chapter may be found in the several works of Balas [6,7], and Jeroslow [23,25].

Chapter II

BASIC CONCEPTS AND PRINCIPLES

2.1 Introduction

In this chapter, we will lead the reader to the most important and fundamental results in disjunctive programming. In order to enable the reader to appreciate the subject matter and to gain better insight into it, we will develop these results from first principles through well known facts. Toward this end, let us commence our discussion with the following well known concept.

2.2 Surrogate Constraints

Let us consider the following constraint set

$$S_1 = \{x: \sum_{j \in N} a_{ij} x_j \geq b_i \text{ for each } i \in Q_1, x \geq 0\} \tag{2.1}$$

where $N = \{1, \ldots, n\}$ is index set for the x-variables and Q_1 is an index set for the linear constraints in S_1 aside from the nonnegativity restrictions. Now let us multiply each of these linear constraints by corresponding nonnegative parameters λ_i, $i \in Q_1$. Then clearly, $x \in S_1$ implies that

$$\sum_{j \in N} \lambda_i a_{ij} x_j \geq \lambda_i b_i \text{ for each } i \in Q_1 \tag{2.2}$$

By simply summing up the constraints (2.2), the following well known result is easily established.

Lemma 2.1

Let S_1 be the constraint set of Equation (2.1). Then $x \in S_1$ implies that

$$\sum_{j \in N} \left\{ \sum_{i \in Q_1} \lambda_i a_{ij} \right\} x_j \geq \sum_{i \in Q_1} \lambda_i b_i \tag{2.3}$$

for any set of nonnegative parameters λ_i, $i \in Q_1$.

Let us consider the converse of Lemma 2.1. In doing so, we are addressing the following question. Suppose that we are given an inequality $\pi x \geq \pi_0$ which is implied by the constraint set S_1. That is, $x \in S$ implies $\pi x \geq \pi_0$. Then, does there exist a surrogate constraint of the form (2.3) obtained through suitable parameters $\lambda_i \geq 0$, $i \in Q_1$ such that this surrogate constraint uniformly dominates the given inequality? The answer is yes. We are able to specify parameters $\lambda_i \geq 0$, $i \in Q_1$ such that if x satisfies (2.3) using these parameters $\lambda_i \geq 0$, $i \in Q_1$, then x must satisfy $\pi x \geq \pi_0$. We establish this result below and then illustrate it through an example.

Lemma 2.2

Let $\pi x \geq \pi_0$ be any inequality implied by S_1 of Equation (2.1) and suppose that S_1 is consistent. Then, there exists a set of nonnegative multipliers λ_i, $i \in Q_1$ such that

$$\sum_{i \in Q_1} \lambda_i a_{ij} \leq \pi_j \text{ for each } j \in N \text{ and } \pi_0 \leq \sum_{i \in Q_1} \lambda_i b_i \qquad (2.4)$$

Proof. Consider the following linear program P and its dual D

$$\underline{P}: \qquad \text{minimize} \quad \left\{ \sum_{j \in N} \pi_j x_j : \sum_{j \in N} a_{ij} x_j \geq b_i, \ i \in Q_i, \ x \geq 0 \right\}$$

$$\equiv \text{minimize} \left\{ \sum_{j \in N} \pi_j x_j \right\}$$
$$\underset{x \in S_1}{}$$

$$\underline{D}: \qquad \text{maximize} \quad \left\{ \sum_{i \in Q_1} \lambda_i b_i : \sum_{i \in Q_1} \lambda_i a_{ij} \leq \pi_j, \ j \in N, \ \lambda_i \geq 0 \right\}$$

$$\text{for each } i \in Q_1$$

Now, since $x \in S_1$ implies $\sum_{j \in N} \pi_j x_j \geq \pi_0$, the primal problem is bounded below by π_0 and hence the feasible region of D is non-empty. Further, since S_1 is consistent, there exists an optimal solution to Problem D. It is easy to see now that the required result holds for any set of dual optimal variables λ_i, $i \in Q_1$. This completes the proof.

Let us illustrate the above result with an example. Consider

$$S_1 = \{x: \ x_1 + 2x_2 \geq 2, \ 3x_1 + x_2 \geq 3, \ x_1, x_2 \geq 0\}$$

14

Now $x_1 + 2x_2 \geq 2$, x_1, $x_2 \geq 0$ imply that $2x_1 + 2x_2 \geq 2$ or that $x_1 + x_2 \geq 1$. Alternatively, $3x_1 + x_2 \geq 3$, x_1, $x_2 \geq 0$ imply that $3x_1 + 3x_2 \geq 3$ or that $x_1 + x_2 \geq 1$. Hence, the inequality $x_1 + x_2 \geq 1$ is implied by S_1. Can we find a surrogate constraint which uniformly dominates this constraint? For this purpose, we consider Problem D in the proof of Lemma 2.2, namely,

$$D: \qquad \text{maximize} \quad 2\lambda_1 + 3\lambda_2$$
$$\text{subject to} \quad \lambda_1 + 3\lambda_2 \leq 1$$
$$2\lambda_1 + \lambda_2 \leq 1$$
$$\lambda_1, \lambda_2 \geq 0$$

One may readily verify that $\lambda_1 = \frac{2}{5}$, $\lambda_2 = \frac{1}{5}$ solves this problem. The surrogate constraint resulting from this is $x_1 + x_2 \geq \frac{7}{5}$. It is also interesting to note that this was a unique optimal solution to Problem D above. Thus, in this case what Problem D essentially did was to translate the cutting plane $x_1 + x_2 \geq 1$ parallel to itself until it supported the feasible region S_1. This is illustrated in Figure 2.1 below.

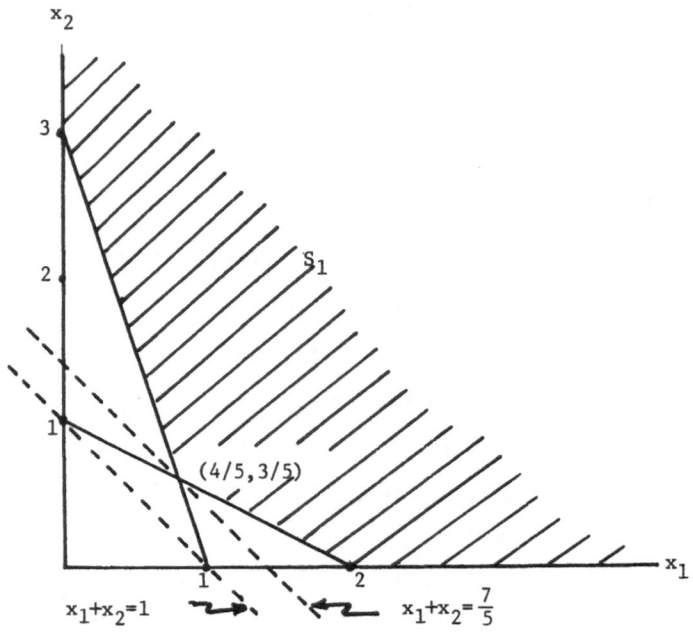

Figure 2.1. Dominance of a Surrogate Constraint

In fact, if $\pi x = \pi_0$ does not support S_1 and $\pi x \geq \pi_0$ is a valid inequality for S_1, then clearly, $x \in S_1$ implies $\pi x > \pi_0$. Thus, if one solves the problem P defined in the proof of Lemma 2.2 and obtains therefrom $\hat{\pi}_0 = \underset{x \in S_1}{\text{minimum}} (\pi x)$ then $\hat{\pi}_0 > \pi_0$ and hence, $\pi x \geq \hat{\pi}_0$ strictly dominates $\pi x \geq \pi_0$. Moreover, $\pi x \geq \hat{\pi}_0$ is a valid inequality for S_1. Thus, Lemma 2.2 would then yield a surrogate constraint which uniformly dominates $\pi x \geq \pi_0$. In other words, non-dominated surrogate constraints support S_1.

2.3 Pointwise-Supremal Cuts

We now proceed to recall another well known, pertinent concept. Suppose that the sets S_h of Equation (1.1) are comprised of a single linear constraint and are given by

$$S_h = \left\{ x: \sum_{j \in N} a_{1j}^h x_j \geq b_1^h, \ x \geq 0 \right\} \text{ for each } h \in H \qquad (2.5)$$

We use the above notation so as to be consistent with the case wherein each S_h, $h \in H$ may contain several constraints as introduced later. We are considering at this point the disjunction $x \in \underset{h \in H}{\bigcup} S_h$.

For example, let $S_1 = \{x: \ 2x_1 - 3x_2 \geq 5, \ x \geq 0\}$ and $S_2 = \{x: \ x_1 + 3x_2 \geq 4, \ x \geq 0\}$ and consider the statement that x satisfies S_1 or S_2. Then, it is a well known fact that x must satisfy $[\max\{2,1\}]x_1 + [\max\{-3,3\}]x_2 \geq \min\{5,4\}$ or $2x_1 + 3x_2 \geq 4$. Thus $2x_1 + 3x_2 \geq 4$ is a valid inequality for the disjunction $x \in \underset{h \in H}{\bigcup} S_h$ since it is implied by each of S_1 and S_2. This fact is generalized and formalized below. First consider the following definition. Then, Lemma 2.3 establishes the required result.

Definition

Consider a set of vectors $\{v^h: \ h \in H\}$ where for each $h \in H$, $v^h = v_1^h, \ldots, v_n^h$. Then, the **pointwise supremum** of this set of vectors, denoted by $\underset{h \in H}{\sup} v^h$, is a vector $v = (v_1, \ldots, v_n)$ with components

$$v_j = \underset{h \in H}{\text{supremum}} \{v_j^h\} \text{ for } j = 1, \ldots, n$$

In a like manner, we define the <u>pointwise infimum</u> of a set of vectors.

<u>Lemma 2.3</u>

Consider the constraint sets S_h, $h \in H$ as defined in Equation (2.5). Then, for the disjunction $x \in \bigcup_{h \in H} S_h$, the following inequality is valid.

$$\sum_{j \in N} \gamma_j x_j \geq \gamma_0, \text{ where } \gamma = \sup_{h \in H} \{a_1^h\}, \quad \gamma_0 = \inf_{h \in H} \{b_1^h\} \qquad (2.6)$$

  <u>Proof.</u> Consider any $\bar{x} \in \bigcup_{h \in H} S_h$. Hence, there exists an $\hat{h} \in H$ for which

$$\sum_{j \in N} a_{1j}^{\hat{h}} \bar{x}_j \geq b_1^{\hat{h}}, \quad \bar{x} \geq 0$$

This implies that

$$\sum_{j \in N} \gamma_j \bar{x}_j = \sum_{j \in N} \left\{ \sup_{h \in H} a_{1j}^h \right\} \bar{x}_j \geq \sum_{j \in N} a_{1j}^{\hat{h}} \bar{x}_j \geq b_1^{\hat{h}} \geq \inf_{h \in H} \left(b_1^h \right) = \gamma_0$$

and the proof is complete.

We will now put Lemmas (2.1), (2.2) and (2.3) together to show that this leads to the fundamental result of disjunctive programming.

<u>2.4 Basic Disjunctive Cut Principle</u>

Suppose that we have constraint sets of the form (1.1), that is.

$$S_h = \left\{ x: \sum_{j \in N} a_{ij}^h x_j \geq b_i^h \text{ for each } i \in Q_h, x \geq 0 \right\}, \quad h \in H \qquad (2.7)$$

where Q_h is an appropriate index set for the constraints in S_h, $h \in H$.

Consider the disjunction $x \in \bigcup_{h \in H} S_h$. Let us now use our discussion in the two proceeding sections to derive valid inequalities for this disjunction.

First of all, note that when each S_h, $h \in H$ has only a single linear constraint, then from Section 2.3 we are able to derive valid inequalities for the disjunction $x \in \bigcup_{h \in H} S_h$. Hence, let us use surrogate constraints to transfrom the given sets S_h into singleton constraint sets and then use the concepts of Section 2.3.

More specifically, let λ_i^h, $i \in Q_h$, $h \in H$ be any set of nonnegative parameters. For each $h \in H$, let us use the corresponding multipliers λ_i^h, $i \in Q_h$, and
from the surrogate constraints

$$\sum_{j \in N} \left\{ \sum_{i \in Q_h} \lambda_i^h a_{ij}^h \right\} x_j \geq \sum_{i \in Q_h} \lambda_i^h b_i^h \quad \text{for each } h \in H \qquad (2.8)$$

Next, let us define sets $\hat{S}_h$, $h \in H$ as follows

$$\hat{S}_h = \{x: \text{ Equation (2.8) is satisfied, } x \geq 0\} \qquad (2.9)$$

Now clearly, $x \in S_h$ implies that $x \in \hat{S}_h$. Hence, the disjunction $x \in \bigcup_{h \in H} S_h$
may be replaced by the (weaker) disjunction $x \in \bigcup_{h \in H} \hat{S}_h$. But then, from
Lemma (2.3), valid inequalities for the latter disjunction are of the form

$$\sum_{j \in N} \left[\sup_{h \in H} \left\{ \sum_{i \in Q_h} \lambda_i^h a_{ij}^h \right\} \right] x_j \geq \inf_{h \in H} \left\{ \sum_{i \in Q_h} \lambda_i^h b_i^h \right\} \qquad (2.10)$$

This result is known as the forward part of the Basic Disjunctive Cut Principle.
To arrive at the converse statement, consider any valid inequality $\sum_{j \in N} \pi_j x_j \geq \pi_0$
implied by $x \in \bigcup_{h \in H} S_h$, and assume that each S_h is consistent. Thus, since
$x \in S_h$ implies $x \in \bigcup_{h \in H} S_h$, then $\pi x \geq \pi_0$ is a valid inequality for each S_h, $h \in H$.
Now, applying Lemma 2.2 for each $h \in H$, we may hence assert that there exist
nonnegative parameters λ_i^h, $i \in Q_h$ such that

$$\left\{ \sum_{i \in Q_h} \lambda_i^h a_{ij}^h \leq \pi_j \text{ for each } j \in N \text{ and } \pi_0 \leq \sum_{i \in Q_h} \lambda_i^h b_i^h \right\} \text{ for each } h \in H$$

This in turn implies that

$$\sup_{h \in H} \left[\sum_{i \in Q_h} \lambda_i^h a_{ij}^h \right] \leq \pi_j \text{ for each } j \in N \text{ and } \pi_0 \leq \inf_{h \in H} \left[\sum_{i \in Q_h} \lambda_i^h b_i^h \right] \qquad (2.11)$$

This result is known as the reverse part of the Basic Disjunctive Cut Principle.
Hence, the forward part gives us a set of valid inequalities for the disjunction
$x \in \bigcup_{h \in H} S_h$, one for each choice of λ_i^h, $i \in Q_h$, $h \in H$. The reverse part then

asserts that if each S_h is consistent, then any valid inequality may be uniformly dominated by a disjunctive cut of the type (2.10). These results are stated formally below.

Theorem 2.1 (Basic Disjunctive Cut Principle)

Let S_h, $h \in H$ be constraint sets given by Equation (1.1). Here, $|H|$ may or may not be finite. Suppose that at least one of the linear inequality systems S_h, $h \in H$ must hold. Then, for any choice of nonnegative vectors $\lambda^h = (\lambda^h_i, i \in Q_h)$, the inequality

$$\left[\sup_{h \in H} (\lambda^h)^t A^h \right] x \geq \inf_{h \in H} (\lambda^h)^t b^h$$

is a valid disjunctive cut, where the superscript t denotes the transpose operation.

Furthermore, if every system S_h is consistent, then for any valid inequality $\sum_{j \in N} \pi_j x_j \geq \pi_0$, there exist nonnegative vectors λ^h, $h \in H$, such that $\pi_0 \leq \inf_{h \in H} (\lambda^h)^t b^h$ and for each $j \in N$, the jth component of $\sup_{h \in H}(\lambda^h)^t A^h$ does not exceed π_j.

Thus far, we have demonstrated that (2.10) yields valid inequalities with no mention being made regarding the selection of values for the parameters λ^h_i, $i \in Q_h$, $h \in H$. This is the subject matter of the next chapter.

2.5 Notes and References

The basic disjunctive cut discussed in Section 2.4 is due to Balas, Glover, and Jeroslow. The forward part appears in Balas [4,6] and the converse in Jeroslow [25]. The same result in a different setting was given by Glover [18,19].

Chapter III

GENERATION OF DEEP CUTS USING THE FUNDAMENTAL
DISJUNCTIVE INEQUALITY

3.1 Introduction

Recall from Chapter I that our motivation in using disjunctive programming

methods is to aid us in solving nonconvex problems of the type

$$\text{DP:} \qquad\qquad \text{minimize} \quad f(x)$$

$$\text{subject to} \quad x \in X \qquad\qquad (3.1)$$

$$x \in \bigcup_{h \in H} S_h \qquad\qquad (3.2)$$

where $f: R^n \to R$ is lower semicontinuous, X is a closed subset of the nonnegative

orthant of R^n and each S_h, $h \in H$ is given by Equation (1.1).

Adopting a relaxation strategy to solve Problem DP suppose we relax con-

straint (3.2). If a solution to the resulting problem is feasible to (3.2), then

it solves Problem DP. Otherwise, we have a point infeasible to the disjunction

(3.2). We thus derive a cut which is valid in the sense that it deletes the

current point, but deletes no point satisfying (3.2). We add this inequality to

the relaxed problem and update the current solution. Thus, at any stage, we

solve the problem to minimize $f(x)$ subject to $x \in X$ and x satisfies the linear

disjunctive inequalities or cuts generated thus far. The procedure terminates

when a solution to such a problem satisfies (3.2).

Now in Chapter II we demonstrated that (2.10) defines valid cuts for the

disjunctive statement (3.2). Given the current point infeasible to (3.2), we

now address the question of selecting nonnegative values for the parameters

λ_i^h, $i \in Q_h$, $h \in H$ in the inequality (2.10) so as to derive a deep disjunctive cut.

We will be devoting our attention to the following two disjunctions titled DC1 and

DC2. We remark that numerous disjunctive statements can be cast in the format of

DC1 or DC2.

<u>DC1</u>:

Suppose that each system S_h is comprised of a single linear inequality, that is, let

$$S_h = \left\{ x: \sum_{j=1}^{n} a_{1j}^{h} x_j \geq b_1^{h}, \quad x \geq 0 \right\} \quad \text{for } h \in H = \{1, \ldots, \hat{h}\} \qquad (3.3)$$

where we assume that $\hat{h} = |H| < \infty$ and that each inequality in S_h, $h \in H$ is stated with the origin as the current point at which the disjunctive cut is being generated. Then, the disjunctive statement DC1 is that at least one of the sets S_h, $h \in H$ must be satisfied. Since the current point (origin) does not satisfy this disjunction, we must have $b_i^{h} > 0$ for each $h \in H$. Further, we will assume, without loss of generality, that for each $h \in H$, $a_{1j}^{h} > 0$ for some $j \in \{1, \ldots, n\}$ or else, S_h is inconsistent and we may disregard it.

<u>DC2</u>:

Suppose each system S_h is comprised of a set of linear inequalitites, that is, let

$$S_h = \left\{ x: \sum_{j=1}^{n} a_{ij}^{h} x_j \geq b_i^{h} \text{ for each } i \in Q_h, \quad x \geq 0 \right\} \quad \text{for } h \in H = \{1, \ldots, \hat{h}\} \qquad (3.4)$$

where Q_h, $h \in H$ are appropriate constraint index sets. Again, we assume that $\hat{h} = |H| < \infty$ and that the representation in (3.4) is with respect to the current point as the origin. Then, the disjunctive statement DC2 is that at least one of the sets S_h, $h \in H$ must be satisfied. Although it is not necessary here for $b_i^{h} > 0$ for all $i \in Q_h$ one may still state a valid disjunction by deleting all constraints with $b_i^{h} \leq 0$, $i \in Q_h$ from each set S_h, $h \in H$. Clearly a valid cut for the relaxed constraint set is valid for the original constraint set. We will thus obtain a cut which possibly is not as strong as may be derived from the original constraints. To aid in our development, we will therefore assume henceforth that $b_i^{h} > 0$, $i \in Q_h$, $h \in H$. Figure 3.1 below illustrates the possible weakening of the cuts derived by such a deletion of constraints. Observe that since a valid cut defines a closed half-space which contains $\underset{h \in H}{U} S_h$, this half space must also contain the

the closure of the convex hull of $\bigcup_{h \in H} S_h$. Since the closure of the convex hull of the union of the sets S_h, $h \in H$ resulting after the deletion of the constraints as above contains the closure of the convex hull of the union of the original sets S_h, $h \in H$, the family of valid cuts derived by the new disjunction are a subset of those that are valid for the original disjunction. Incidentally, one may also note that facets of the closure of the convex hull of feasible points are desirable deep cuts.

Before proceeding with our analysis, let us briefly comment on the need for deep cuts. Although intuitively desirable, it is not always necessary to seek a deepest cut. For example, if one is using cutting planes to implicitly search a feasible region of discrete points, then all cuts which delete the same subset of this discrete region may be equally attractive irrespective of their depth relative to the convex hull of this discrete region. On the other hand, if one is confronted with the problem of iteratively exhausting a feasible region which is not finite, then indeed deep cuts are meaningful and desirable.

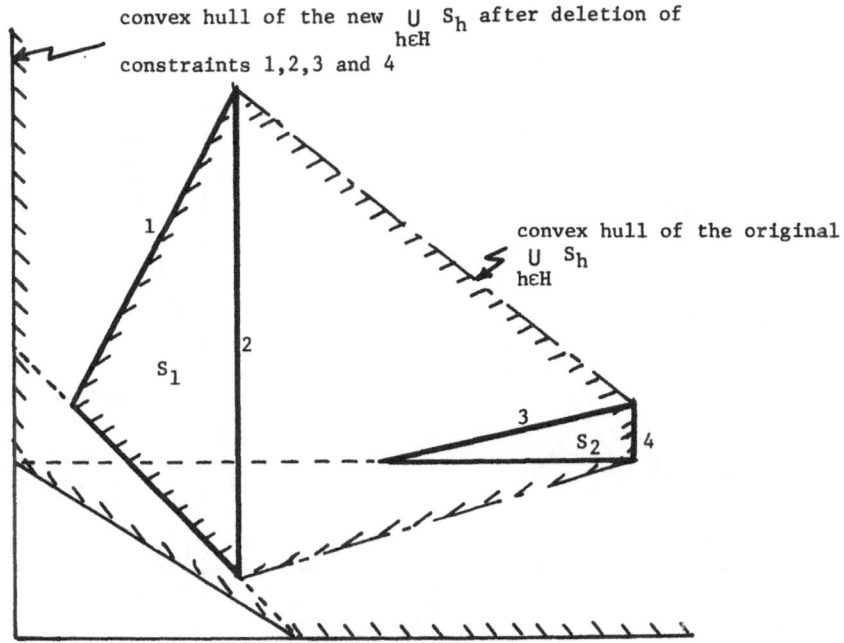

Figure 3.1. Formulation of the Disjunction DC2

3.2 Defining Suitable Criteria for Evaluating the Depth of a Cut

In this section, we will lay the foundation for the concepts we propose to use in deriving deep cuts. Specifically, we will explore the following two criteria for deriving a deep cut:

(i) Maximize the euclidean distance between the origin and the nonnegative region feasible to the cutting plane

(ii) Maximize the rectilinear distance between the origin and the nonnegative region feasible to the cutting plane.

Let us briefly discuss the choice of these criteria. Referring to Figure 3.2(a) below, one may observe that simply attempting to maximize the euclidean distance from the origin to the cut can favor a weaker cut over stronger cuts. However, since one is only interested in the subset of the nonnegative orthant feasible to the cuts, the choice of criterion (i) above avoids such anamolies. Of course, as Figure 3.2(b) indicates, it is possible for this criterion to be unable to recognize dominance and treat one cut and another one which dominates it as alternative optimal cuts.

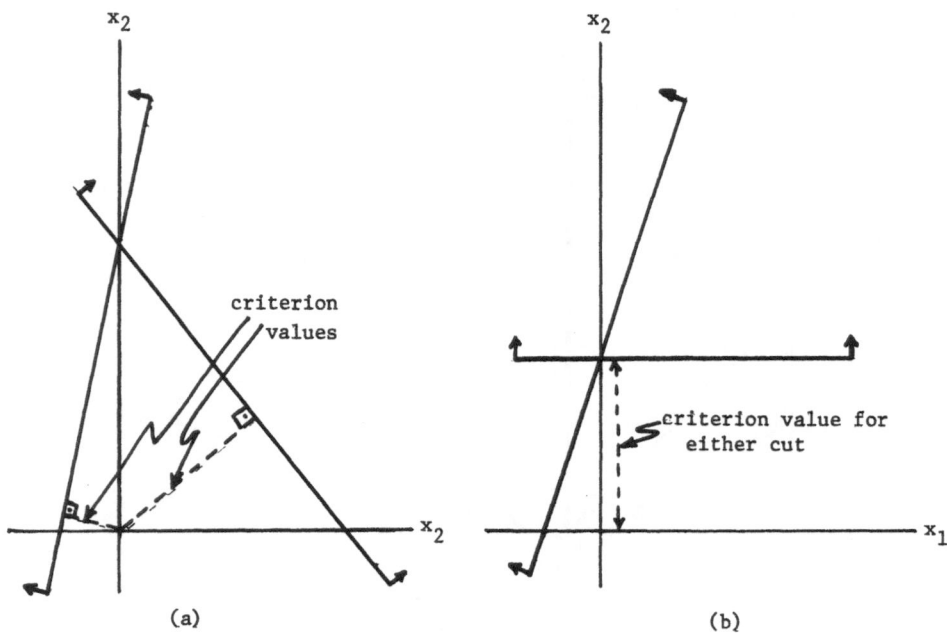

Figure 3.2. Recognition of Dominance

Let us now proceed to characterize the euclidean distance from the origin to the nonnegative region feasible to a cut

$$\sum_{j=1}^{n} z_j x_j \geq z_0, \text{ where } z_0 > 0, \ z_j > 0 \text{ for some } j \varepsilon \{1,\ldots,n\} \tag{3.5}$$

The required distance is clearly given by

$$\theta_e = \text{minimum } \{\|x\| : \sum_{j=1}^{n} z_j x_j \geq z_0, \ x \geq 0\} \tag{3.6}$$

where $\|x\| = \sqrt{\sum_{j=1}^{n} x_j^2}$. Consider the following result

Lemma 3.1

Let θ_e be defined by Equation (3.5) and (3.6). Then

$$\theta_e = \frac{z_0}{\|y\|} \tag{3.7}$$

where,

$$y = (y_1,\ldots,y_n), \ y_j = \text{maximum } \{0, z_j\}, \ j=1,\ldots,n \tag{3.8}$$

Proof. Note that the solution $x^* = \left(\dfrac{z_0}{\|y\|^2}\right) y$ is feasible to the problem in (3.6) with $\|x^*\| = \dfrac{z_0}{\|y\|}$. Moreover, for any x feasible to (3.6), we have,

$z_0 \leq \sum_{j=1}^{n} z_j x_j \leq \sum_{j=1}^{n} y_j x_j \leq \|y\| \ \|x\|$, or that, $\|x\| \geq \dfrac{z_0}{\|y\|}$. This completes the proof.

Now, let us consider the second criterion. The motivation for this criterion is similar to that for the first criterion and moreover, as we shall see below, the use of this criterion has intuitive appeal. First of all, given a cut (3.5), let us characterize the rectilinear distance from the origin to the nonnegative region feasible to this cut. This distance is given by

$$\theta_r = \text{minimum } \{|x|: \sum_{j=1}^{n} z_j x_j \geq z_0, \ x \geq 0\} \tag{3.9}$$

where $|x| = \sum_{j=1}^{n} |x_j|$. Consider the following result.

Lemma 3.2

Let θ_r be defined by Equations (3.5) and (3.9). Then,

$$\theta_r = \frac{z_0}{z_m} \quad \text{where } z_m = \underset{j=1,\ldots,n}{\text{maximum}} z_j \tag{3.10}$$

Proof. Note that the solution $x* = (0,\ldots,\frac{z_0}{z_m},\ldots,0)$, with the mth component being non-zero, is feasible to the problem in (3.9) with $|x*| = \frac{z_0}{z_m}$. Moreover, for any x feasible to (3.9), we have,

$$\frac{z_0}{z_m} \leq \sum_{j=1}^{n} \frac{z_j}{z_m} x_j \leq \sum_{j=1}^{n} x_j = |x|$$

This completes the proof.

Note from Equation (3.10) that the objective of maximizing θ_r is equivalent to finding a cut which maximizes the smallest positive intercept made on any axis. Hence, the intuitive appeal of this criterion.

3.3 Deriving Deep Cuts for DC1

It is very encouraging to note that for the disjunction DC1 we are able to derive a cut which not only simultaneously satisfies both the criterion of Section 3.2, but which is also a facet of the set S defined by (3.11) below.

$$S = \text{closure convex hull of } \bigcup_{h \in H} S_h \tag{3.11}$$

This is a powerful statement since all valid inequalities are given through (2.10) and none of these can strictly dominate a facet of S.

We will find it more convenient to state our results if we normalize the linear inequalities (3.3) by dividing through by their respective, positive, right-hand-sides. Hence, let us assume without loss of generality that

$$S_h = \left\{ x: \sum_{j=1}^{n} a_{1j}^h x_j \geq 1, \ x \geq 0 \right\} \quad \text{for } h \epsilon H = \{1, \ldots, \hat{h}\} \qquad (3.12)$$

Then the application of Theorem 2.1 to the disjunction DC1 yields valid cuts of the form:

$$\sum_{j=1}^{n} \left\{ \max_{h \epsilon H} \lambda_1^h a_{1j}^h \right\} x_j \geq \min_{h \epsilon H} \{\lambda_1^h\} \qquad (3.13)$$

where λ_1^h, $h \epsilon H$ are nonnegative scalars. Again, there is no loss of generality in assuming that

$$\sum_{h \epsilon H} \lambda_1^h = 1, \ \lambda_1^h \geq 0, \ h \epsilon H = \{1, \ldots, \hat{h}\} \qquad (3.14)$$

since we will not allow all λ_1^h, $h \epsilon H$ to be zero. This is equivalent to normalizing (3.13) be dividing through by $\sum_{h \epsilon H} \lambda_1^h$.

Theorem 3.1 below derives two cuts of the type (3.13), both of which simultaneously achieve the two criteria of the foregoing section. However, the second cut uniformly dominates the first cut. In fact, no cut can strictly dominate the second cut since it is shown to be a facet of S defined by (3.11).

Theorem 3.1

Consider the disjunctive statement DC1 where S_h is defined by (3.12) and is assumed to be consistent for each $h \epsilon H$. Then the following results hold

(a) Both the criteria of Section 3.2 are satisfied by letting

$$\lambda_1^h = 1/\hat{h} = \lambda_1^{h*}, \ \text{say, for } h \epsilon H \qquad (3.15)$$

in inequality (3.13) to obtain the cut

$$\sum_{j=1}^{n} a_{1j}^* x_j \geq 1, \ \text{where } a_{1j}^* = \max_{h \epsilon H} a_{1j}^h, \ j=1, \ldots, n \qquad (3.16)$$

(b) Further, defining

$$\gamma_1^h = \underset{j:a_{1j}^h>0}{\text{minimum}} \{a_{1j}^*/a_{1j}^h\} > 0, \; h \, \varepsilon \, H \tag{3.17}$$

and letting

$$\lambda_1^h = \gamma_1^h \Big/ \sum_{p \varepsilon H} \gamma_1^p = \lambda_1^{h**}, \text{ say, for } h \, \varepsilon \, H \tag{3.18}$$

in inequality (3.13), we obtain a cut of the form

$$\sum_{j=1}^{n} a_{1j}^{**} x_j \geq 1 \tag{3.19}$$

which again satisfies both the criteria of Section 3.2.

(c) The cut (3.19) uniformly dominates the cut (3.16); in fact,

$$a_{1j}^{**} \begin{cases} = a_{1j}^* \text{ if } a_{1j}^* > 0 \\ \leq a_{1j}^* \text{ if } a_{1j}^* \leq 0 \end{cases}, \quad j=1,\ldots,n \tag{3.20}$$

(d) The cut (3.19) is a facet of the set S of Equation (3.11).

Proof.

(a) Clearly, $\lambda_1^h = 1/\hat{h}$, $h \, \varepsilon \, H$ leads to the cut (3.16) from (3.13). Now consider the euclidean distance criterion of maximizing θ_e (or θ_e^2) of Equation (3.7). For cut (3.16), the value of θ_e^2 is given by

$$(\theta_e^*)^2 = 1 \Big/ \sum_{j=1}^{n} (y_j^*)^2 > 0 \text{ where } y_j^* = \max\{0, a_{1j}^*\}, \; j=1,\ldots,n \tag{3.21}$$

Now, for any choice λ_1^h, $h \, \varepsilon \, H$,

$$\theta_e^2 = \left[\underset{h \varepsilon H}{\min}(\lambda_1^h) \right]^2 \Big/ \sum_{j=1}^{n} y_j^2 = (\lambda_1^p)^2 \Big/ \sum_{j=1}^{n} y_j^2, \text{ say,} \tag{3.22}$$

where $y_j = \max\{0, \underset{h \varepsilon H}{\max} \; \lambda_1^h a_{1j}^h\}$. If $\lambda_1^p = 0$, then $\theta_e = 0$ and noting (3.21), such a choice of parameters λ_1^h, $h \, \varepsilon \, H$ is suboptimal. Hence, $\lambda_1^p > 0$, whence (3.22) becomes

$\theta_e^2 = 1 \Big/ \sum\limits_{j=1}^{n} \left(\dfrac{y_j}{\lambda_1^p}\right)^2$. But since $(\lambda_1^h/\lambda_1^p) \geq 1$ for each $h \in H$, we get

$$y_j/\lambda_1^p = \max\{0, \max\limits_{h \in H} \dfrac{\lambda_1^h}{\lambda_1^p} \, a_{1j}^h\} \geq \max\{0, \max\limits_{h \in H} a_{1j}^h\} = y_j^*$$

Thus $\theta_e^2 \leq (\theta_e^*)^2$ or that the first criterion is satisfied.

Now consider the maximization of θ_r of Equations (3.9), (3.10). For the choice (3.15), the value of θ_r is given by

$$\theta_r^* = \dfrac{1}{\max\limits_{j} a_{1j}^*} > 0 \qquad\qquad (3.23)$$

Now, for any choice λ_1^h, $h \in H$, from Equations (3.10), (3.6) we get

$$\theta_r = \left(\min\limits_{h \in H} \lambda_1^h\right) \Big/ \left(\max\limits_{j} \max\limits_{h \in H} \lambda_1^h a_{1j}^h\right) = \lambda_1^p \Big/ \max\limits_{j} \max\limits_{h \in H} \lambda_1^h a_{1j}^h, \text{ say.}$$

As before, $\lambda_1^p = 0$ implies a value of θ_r inferior to θ_r^*. Thus, assume $\lambda_1^p > 0$.

Then, $\theta_r = 1 \Big/ \max\limits_{j} \max\limits_{h \in H} \left(\dfrac{\lambda_1^h}{\lambda_1^p}\right) a_{1j}^h$. But $(\lambda_1^h/\lambda_1^p) \geq 1$ for each $h \in H$ and in evaluating θ_r, we are interested only in those $j \in \{1,\ldots,n\}$ for which $a_{1j}^h > 0$ for some $h \in H$.

Thus $\theta_r \leq 1/\max\limits_{j} \max\limits_{h \in H} a_{1j}^h = \theta_r^*$, or that the second criterion is also satisfied.
This proves part (a).

(b) and (c). First of all, let us consider the values taken by γ_1^h, $h \in H$. Note from the assumption of consistency that γ_1^h, $h \in H$ are well defined. From (3.16), (3.17), we must have $\gamma_1^h \geq 1$ for each $h \in H$. Moreover, if we define from (3.16)

$$H^* = \{h \in H: a_{1k}^h = a_{1k}^* > 0 \text{ for some } k \in \{1,\ldots,n\}\} \qquad\qquad (3.24)$$

then clearly $H^* \neq \{\phi\}$ and for $h \in H^*$, Equation (3.17) implies $\gamma_1^h \leq 1$. Thus,

$$\gamma_1^h \begin{cases} = 1 & \text{for } h \in H^* \\ > 1 & \text{for } h \notin H^* \end{cases} \qquad\qquad (3.25)$$

Hence,

$$\min_{h \in H} \gamma_1^h = 1 \tag{3.26}$$

or that, using (3.18) in (3.13) yields a cut of the type (3.19), where,

$$a_{1j}^{**} = \max_{h \in H} a_{1j}^h \gamma_1^h \ , \ j=1,\ldots,n \tag{3.27}$$

Now, let us establish relationship (3.20). Note from (3.16) that if $a_{1j}^* \leq 0$, then $a_{1j}^h \leq 0$ for each $h \in H$ and hence, using (3.25), (3.27), we get that (3.20) holds. Next, consider $a_{1j}^* > 0$ for some $j \in \{1,\ldots,n\}$. From (3.24), (3.25), (3.27), we get

$$a_{1j}^{**} = \max\{\max_{h \in H} \cdot a_{1j}^h, \ \max_{\substack{h \in H^* \\ a_{1j}^h > 0}} a_{1j}^h \gamma_1^h\} \tag{3.28}$$

where we have not considered $h \notin H^*$ with $a_{1j}^h \leq 0$ since $a_{1j}^{**} > 0$. But for $h \notin H^*$ with $a_{1j}^h > 0$, we get from (3.16), (3.17)

$$a_{1j}^h \gamma_1^h = a_{1j}^h \left[\min_{k:a_{1k}^h > 0} \left\{ \frac{\max_{r \in H} a_{1k}^r}{a_{1k}^h} \right\} \right] \leq a_{1j}^h \left\{ \frac{\max_{r \in H} a_{1j}^r}{a_{1j}^h} \right\} = \max_{r \in H} a_{1j}^r \tag{3.29}$$

Using (3.29) in (3.28) yields $a_{1j}^{**} = a_{1j}^*$, which establishes (3.20).

Finally, we show that (3.19) satisfies both the criteria of Section 3.2. This part follows immediately from (3.20) by noting that the cut (3.16) yields $\theta_e = \theta_e^*$ of (3.21) and $\theta_r = \theta_r^*$ of (3.23). This completes the proofs of parts (b) and (c).

(d) Note that since (3.19) is valid, any $x \in S$ satisfies (3.19). Hence, in order to show that (3.19) defines a facet of S, it is sufficient to identify n affinely independent points of S which satisfy (3.19) as an equality, since clearly, S is of dimension n. Define

$$J_1 = \{j \in \{1,\ldots,n\}: a_{1j}^{**} > 0\} \text{ and let } J_2 = \{1,\ldots,n\} - J_1 \tag{3.30}$$

Consider any $p\epsilon J_1$, and let

$$e_p = (0,\ldots,\frac{1}{a_{1p}^{**}},\ldots,0), \quad p\epsilon J_1 \qquad (3.31)$$

have the non-zero term in the pth position. Now, since $p\epsilon J_1$, (3.20) yields

$$a_{1p}^{**} = a_{1p}^{*} = \max_{h\epsilon H} a_{1p}^{h} = a_{1p}^{hp} , \text{ say.}$$

Hence, $e_p \epsilon S_{h_p}$ and so, $e_p\epsilon S$ and moreover, e_p satisfies (3.19) as an equality. Thus, $e_p, p\epsilon J_1$ qualify as $|J_1|$ of the n affinely independent points we are seeking. Now consider a $q\epsilon J_2$. Let us show that there exists an S_{h_q} satisfying

$$\gamma_1^{hq} a_{1p}^{hq} = a_{1p}^{**} \text{ for some } p\epsilon J_1$$

and $\hspace{10cm} (3.32)$

$$\gamma_1^{hq} a_{1q}^{hq} = a_{1q}^{**}$$

From Equation (3.27), we get $a_{1q}^{**} = \max_{h\epsilon H} a_{1q}^{h}\gamma_1^{h} = a_{1q}^{hq}\gamma_1^{hq}$, say. Then for this $h_q\epsilon H$,

Equation (3.17) yields $\gamma_1^{hq} = \underset{j:a_{1j}^{hq}>0}{\text{minimum}} \{a_{1j}^{*}/a_{1j}^{hq}\} = a_{1p}^{*}/a_{1p}^{hq}$, say. Or, using (3.20)

$\gamma_1^{hq} a_{1p}^{hq} = a_{1p}^{*} = a_{1p}^{**} > 0$. Thus (3.32) holds. For convenience, let us rewrite the set S_{h_q} below as

$$S_{h_q} = \{x: a_{1p}^{hq} x_p + a_{1q}^{hq} x_q + \sum_{j\neq p,q} a_{1j}^{hq} x_j \geq 1, x \geq 0\} \qquad (3.33)$$

Now, consider the direction

$$d_q = \begin{cases} (0,\ldots,\frac{1}{a_{1p}^{**}},\ldots, -\frac{1}{a_{1q}^{**}}, \ldots,0) & \text{if } a_{1q}^{**} < 0 \\ \\ (0,\ldots, \quad 0 \quad,\ldots, \quad\quad \Delta , \ldots,0) & \text{if } a_{1q}^{**} = 0 \end{cases} \qquad (3.34)$$

where $\Delta > 0$. Let us shown that d_q is a direction for S_{h_q}. Clearly, if $a_{1q}^{**} = 0$, then from (3.32) $a_{1q}^{hq} = 0$ and thus (3.33) establishes (3.34). Further, if $a_{1q}^{**} < 0$ then one may easily verify from (3.32), (3.33), (3.34) that

$$\hat{e}_p = (0,\ldots,\gamma_1^{hq}/a_{1p}^{**},\ldots,0) \ \varepsilon \ S_{h_q} \text{ and } \hat{e}_p + \delta[\gamma_1^{hq} \ d_q] \ \varepsilon \ S_{h_q} \text{ for each } \delta \geq 0$$

where $\hat{e}_p$ has the non-zero term at position p. Thus, d_q is a direction for S_{h_q}. It can be easily shown that this implies d_q is a direction for S. Since $e_p = (0,\ldots,\dfrac{1}{a_{1p}^{**}},\ldots,0)$ of Equation (3.31) belongs to S, then so does $(e_p + d_q)$. But $(e_p + d_q)$ clearly satisfies (3.19) as an equality. Hence, we have identified n points of S, which satisfy the cut (3.19) as an equality, of the type

$$\left. \begin{array}{l} e_p = (0,\ldots,\dfrac{1}{a_{1p}^{**}},\ldots,0) \text{ for } p\varepsilon J_1 \\ \\ e_q = d_q + e_p \text{ for some } p\varepsilon J_1, \text{ for each } q\varepsilon J_2 \end{array} \right\} \qquad (3.35)$$

where d_q is given by (3.34). Since these n points are clearly affinely independent, this completes the proof.

Thus, in view of Theorem 3.1, it is "optimal" to derive a cut (3.19) for the disjunction DC1. In generalizing this to disjunction DC2, we find that such an ideal situation no longer exists. Nevertheless, we are able to obtain some meaningful results. But before proceeding to DC2, let us illustrate the above concepts through an example.

Example

Let $H = \{1,2\}$, n=3 and let DC1 be formulated through the sets

$$S_1 = \{x: x_1 + 2x_2 - 4x_3 \geq 1, \ x \geq 0\}, \ S_2 = \{x: \frac{x_1}{2} + \frac{x_2}{3} - 2x_3 \geq 1, \ x \geq 0\}.$$

The cut (3.16), i.e., $\sum a_{1j}^* x_j \geq 1$, is $x_1 + 2x_2 - 2x_3 \geq 1$. From (3.17),

$$\gamma_1^1 = \min \ \{\frac{1}{1}, \frac{2}{2}\} = 1 \text{ and } \gamma_1^2 = \min\{\frac{1}{1/2}, \frac{2}{1/3}\} = 2$$

Thus, through (3.18), or more directly, from (3.27), the cut (3.19), i.e.,

$\sum a_{1j}^{**} x_j \geq 1$ is $x_1 + 2x_2 - 4x_3 \geq 1$. This cut strictly dominates the cut (3.16) in this example, though both have the same values $1/\sqrt{5}$ and $1/2$ respectively for θ_e and θ_r of Equations (3.6) and (3.9) respectively.

3.4 Deriving Deep Cuts for DC2

To begin with, let us make the following interesting observation. Suppose that for convenience, we assume without loss of generality as before, that $b_1^h = 1$, $i \varepsilon Q_h$, $h \varepsilon H$ in Equation (3.3). Thus, for each $h \varepsilon H$, we have the constraint set

$$S_h = \left\{ x: \sum_{j=1}^{n} a_{ij}^h x_j \geq 1, \ i \varepsilon Q_h, \ x \geq 0 \right\} \tag{3.35}$$

Now for each $h \varepsilon H$, let us multiply the constraints of S_h by corresponding scalars $\delta_i^h \geq 0$, $i \varepsilon Q_h$ and add them up to obtain the surrogate constraint

$$\sum_{j=1}^{n} \left\{ \sum_{i \varepsilon Q_h} \delta_i^h a_{ij}^h \right\} x_j \geq \sum_{i \varepsilon Q_h} \delta_i^h, \ h \varepsilon H \tag{3.37}$$

Or, assuming that not all δ_i^h are zero for $i \varepsilon Q_h$, i.e., letting each set S_h, $h \varepsilon H$ govern the cut, (3.37) may be re-written as

$$\sum_{j=1}^{n} \left\{ \sum_{i \varepsilon Q_h} \left[\frac{\delta_i^h}{\left(\sum_{p \varepsilon Q_h} \delta_p^h \right)} \right] a_{1j}^h \right\} x_j \geq 1, \ h \varepsilon H \tag{3.38}$$

Finally, denoting $\delta_i^h \Big/ \sum_{p \varepsilon Q_h} \delta_p^h$ by λ_i^h for $i \varepsilon Q_h$, $h \varepsilon H$, we may write (3.38) as

$$\sum_{j=1}^{n} \left(\sum_{i \varepsilon Q_h} \lambda_i^h a_{ij}^h \right) x_j \geq 1 \text{ for each } h \varepsilon H \tag{3.39}$$

where,

$$\sum_{i \varepsilon Q_h} \lambda_i^h = 1 \text{ for each } h \varepsilon H, \ \lambda_i^h \geq 0 \text{ for } i \varepsilon Q_h, \ h \varepsilon H \tag{3.40}$$

Observe that by surrogating the constraints of (3.36) using parameters λ_i^h, $i\epsilon Q_h$, $h\in H$ satisfying (3.40), we have essentially represented DC2 as DC1 through (3.39). In other words, since $x\epsilon S_h$ implies x satisfies (3.39) for each $h\in H$, then given λ_i^h, $i\epsilon Q_h$, $h\in H$, DC2 implies that at least one of (3.39) must be satisfied. Now, whereas Theorem 2.1 would directly employ (3.37) to derive a cut, since we have normalized (3.37) to obtain (3.39), we know from the previous section that the optimal strategy is to derive a cut (3.19) using inequalities (3.39).

Now let us consider in turn the two criteria of Section 3.2.

3.4.1 Euclidean Distance-Based Criterion

Consider any selection of values for the parameters λ_i^h, $i\epsilon Q_h$, $h\in H$ satisfying (3.40) and let the corresponding disjunction DC1 derived from DC2 be that at least one of (3.39) must hold. Then, Theorem 3.1 tells us through Equations (3.16), (3.21) that the euclidean distance criterion value for the resulting cut (3.19) is

$$\theta_e(\lambda) = 1 \bigg/ \sqrt{\sum_{j=1}^{n} y_j^2} \qquad (3.41)$$

where,

$$y_j = \max\{0, z_j\}, \quad j=1,\ldots,n \qquad (3.42)$$

and

$$z_j = \max_{h\in H} \left\{ \sum_{i\epsilon Q_h} \lambda_i^h a_{ij}^h \right\}, \quad j=1,\ldots,n \qquad (3.43)$$

Thus, the criterion of Section 3.2 seeks to

$$\text{maximize}\{\theta_e(\lambda): \lambda = (\lambda_i^h) \text{ satisfies } (3.40)\} \qquad (3.44)$$

or equivalently, to

$$\text{minimize } \left\{ \sum_{j=1}^{n} y_j^2: (3.40), (3.42), (3.43) \text{ are satisfied} \right\}. \qquad (3.45)$$

It may be easily verified that the problem of (3.45) may be written as

$$\underline{PD_2}: \qquad \text{minimize} \qquad \sum_{j=1}^{n} y_j^2 \qquad\qquad\qquad (3.46)$$

$$\text{subject to} \qquad y_j \geq \sum_{i \in Q_h} \lambda_i^h a_{ij}^h \quad \text{for each } h \varepsilon H$$
$$\text{for each } j=1,\ldots,n \qquad (3.47)$$

$$\sum_{i \in Q_h} \lambda_i^h = 1 \qquad \text{for each } h \varepsilon H \qquad\qquad (3.48)$$

$$\lambda_i^h \geq 0 \quad i \varepsilon Q_h, \; h \varepsilon H \qquad\qquad\qquad (3.49)$$

The equivalence follows by noting that any optimal solution to PD_2 must satisfy (3.42) as an equality. In particular, we have deleted the constraints $y_j \geq 0$, $j=1,\ldots,n$ since for any feasible λ_i^h, $i \varepsilon Q_h$, $h \varepsilon H$, there exists a dominant solution with nonnegative y_j, $j=1,\ldots,n$. This relaxation is simply a matter of convenience in our solution strategy.

Before proposing a solution procedure for Problem PD_2, let us make some pertinent remarks. Note that Problem PD_2 has the purpose of generating parameters λ_i^h, $i \varepsilon Q_h$, $h \varepsilon H$ which are to be used to obtain the surrogate constraints (3.39). Thereafter, the cut that we derive for the disjunction DC_2 is the cut (3.19) obtained from the statement that at least one of (3.39) must hold. Hence, Problem PD_2 attempts to find values for λ_i^h, $i \varepsilon Q_h$, $h \varepsilon H$, such that this resulting cut achieves the euclidean distance criterion.

Problem PD_2 is a convex quadratic program for which the Karush-Kuhn-Tucker conditions are both necessary and sufficient. Several efficient simplex-based quadratic programming procedures are available to solve such a problem. However, these procedures require explicit handling of the potentially large number of constraint in Problem PD_2. On the other hand, the subgradient optimization procedure discussed below takes full advantage of the problem structure. We are first able to write out an almost complete solution to the Karush-Kuhn-Tucker system. We will refer to this as a <u>partial solution</u>. In case we are unable to either actually construct a complete solution or to assert that a feasible completion exists, then through the construction procedure itself, we will show that a sub-gradient direction is available. Moreover, this latter direction is very likely to be a direction of ascent. We therefore propose to move in the negative of this direction and if necessary, project back onto the feasible region. These iterative steps are now repeated at this new point.

3.4.1 Karush-Kuhn-Tucker Systems for PD$_2$ and Its Implications

Letting u_j^h, $h \varepsilon H$, $j=1,\ldots,n$ denote the lagrangian multipliers for constraints (3.47), t_h, $h \varepsilon H$ those for constraints (3.48), and w_i^h, $i \varepsilon Q_h$, $h \varepsilon H$ those for constraints (3.49), we may write the Karush-Kuhn-Tucker optimality conditions as

$$\sum_{h \varepsilon H} u_j^h = 2y_j \qquad j=1,\ldots,n \tag{3.50}$$

$$\sum_{j=1}^{n} u_j^h a_{ij}^h + t_h - w_i^h = 0 \text{ for each } i \varepsilon Q_h, \text{ and for each } h \varepsilon H \tag{3.51}$$

$$u_j^h \left\{ \sum_{i \varepsilon Q_h} \lambda_i^h a_{ij}^h - y_j \right\} = 0 \quad \text{for each } j=1,\ldots,n, \text{ and each } h \varepsilon H \tag{3.52}$$

$$\lambda_i^h w_i^h = 0 \text{ for } i \varepsilon Q_h, \ h \varepsilon H \tag{3.53}$$

$$w_i^h \geq 0 \quad i \varepsilon Q_h, \ h \varepsilon H \tag{3.54}$$

$$u_j^h \geq 0 \quad j=1,\ldots,n, \ h \varepsilon H \tag{3.55}$$

Finally, Equations (3.47), (3.48), (3.49) must also hold.

We will now consider the implications of the above conditions. This will enable us to construct at least a partial solution to these conditions, given particular values of λ_i^h, $i \varepsilon Q_h$, $h \varepsilon H$. First of all, note that Equations (3.42), (3.45) and (3.55) imply that

$$y_j \geq 0 \text{ for each } j=1,\ldots,n \tag{3.56}$$

$$y_j = \max \left\{ 0, \sum_{i \in Q_h} \lambda_i^h a_{ij}^h, \; h \in H \right\} \quad \text{for } j=1,\ldots,n \tag{3.57}$$

Now, having determined values for $y_j = j=1,\ldots,n$, let us define the sets

$$H_j = \begin{cases} \{\Phi\} \text{ if } y_j = 0 \\ \\ \{h \in H: \; y_j = \sum_{i \in Q_h} \lambda_i^h a_{ij}^h > 0\} \end{cases} \quad \text{for } j=1,\ldots,n \tag{3.58}$$

Now, consider the determination of u_j^h, $h \in H$, $j=1,\ldots,n$. Clearly, Equations (3.50), (3.52) and (3.55) along with the definition (3.58) imply that for each $j=1,\ldots,n$

$$u_j^h = 0 \text{ for } h \in H/H_j \text{ and that } \sum_{h \in H_j} u_j^h = 2y_j, \; u_j^h \geq 0 \text{ for each } h \in H_j \tag{3.59}$$

Thus, for any $j \in \{1,\ldots,n\}$, if H_j is either empty or a singleton, the corresponding values for u_j^h, $h \in H$ are uniquely determined. Hence, we have a choice in selecting values for u_j^h, $h \in H_j$ only when $|H_j| \geq 2$ for any $j \in \{1,\ldots,n\}$. Next, multiplying (3.51) by λ_i^h and using (3.53), we obtain

$$\sum_{j=1}^{n} \left[u_j^h \sum_{i \in Q_h} \{\lambda_i^h a_{ij}^h\} \right] + t_h \sum_{i \in Q_h} \lambda_i^h = 0 \text{ for each } h \in H \tag{3.60}$$

Using Equations (3.48), (3.52), this gives us

$$t_h = - \sum_{j=1}^{n} u_j^h y_j \quad \text{for each } h \in H \tag{3.61}$$

Finally, Equations (3.51), (3.61) yield

$$w_i^h = \sum_{j=1}^{n} u_j^h [a_{ij}^h - y_j] \quad \text{for each } i \in Q_h, \; h \in H \tag{3.62}$$

Notice that once the variables u_j^h, $h \in H$, $j=1,\ldots,n$ are fixed to satisfy (3.59), all the variables are uniquely determined. We now show that if the variables w_j^h, $i \epsilon Q_h$, $h \in H$ so determined are nonnegative, we then have a Karush-Kuhn-Tucker solution. Since the objective function of PD$_2$ is convex and the constraints are linear, this solution is also optimal.

Lemma 3.3

Let a primal feasible set of values for λ_i^h, $i \epsilon Q_h$, $h \in H$ be given. Determine values for all variables y_j, u_j^h, t_h, w_i^h using Equations (3.57) through (3.62), selecting an arbitrary solution in the case described in Equation (3.59) if $|H_j| \geq 2$. If $w_i^h \geq 0$, $i \epsilon Q_h$, $h \in H$, then λ_i^h, $i \epsilon Q_h$, $h \in H$ solves Problem PD$_2$.

Proof. By construction Equations (3.47) through (3.52), and (3.55) clearly hold. Thus, noting that in our problem the Kuhn-Tucker conditions are sufficient for optimality, all we need to show is that if $w = (w_i^h) \geq 0$ then (3.53) holds. But from (3.52) and (3.62) for any $h \in H$, we have,

$$\sum_{i \epsilon Q_h} \lambda_i w_i^h = \sum_{i \epsilon Q_h} \lambda_i^h \left\{ \sum_{j=1}^n u_j^h [a_{ij}^h - y_j] \right\} = \sum_{j=1}^n \left\{ u_j^h [\sum_{i \epsilon Q_h} \lambda_i^h a_{ij}^h - y_j] \right\} = 0$$

$$\text{for each } h \epsilon H.$$

Thus, $\lambda_i^h \geq 0$, $w_i^h \geq 0$ $i \epsilon Q_h$, $h \in H$ imply that (3.53) holds and the proof is complete.

The reader may note that in Section 3.4.1(d) we will propose another stronger sufficient condition for a set of variables λ_i^h, $i \epsilon Q_h$, $h \in H$ to be optimal. The development of this condition is based on a subgradient optimization procedure discussed below.

3.4.1(b) Subgradient Optimization Scheme for Problem PD$_2$

For the purpose of this development, let us use (3.57) to rewrite Problem PD$_2$ as follows. First of all define

$$\Lambda = \{\lambda = (\lambda_i^h): \text{constraints (3.48) and (3.49) are satisfied}\} \qquad (3.63)$$

and let $F: \Lambda \to R$ be defined by

$$F(\lambda) = \sum_{j=1}^{n} [\text{maximum } \{0, \sum_{i \epsilon Q_h} \lambda_i^h a_{ij}^h, h \epsilon H\}]^2 \qquad (3.64)$$

Then, Problem PD_2 may be written as

$$\text{minimize } \{F(\lambda): \lambda \epsilon \Lambda\}$$

Note that for each $j=1,\ldots,n$, $g_j(\lambda) = \max \{0, \sum_{i \epsilon Q_h} \lambda_i^h a_{ij}^h, h \epsilon H\}$ is convex and

nonnegative. Thus, $[g_j(\lambda)]^2$ is convex and so $F(\lambda) = \sum_{j=1}^{n} [g_j(\lambda)]^2$ is also convex.

The main thrust of the proposed algorithm is as follows. Having a

solution $\bar{\lambda}$ at any stage, we will attempt to construct a solution to the Karush-

Kuhn-Tucker system using Equations (3.50) through (3.55). If we obtain non-

negative values $\bar{w}_i^h$ for the corresponding variables w_i^h, $i \epsilon Q_h$, $h \epsilon H$, then by Lemma

3.3 above, we terminate. Later in Section 3.4.1(g), we will also use another

sufficient condition to check for termination. If we obtain no indication of

optimality, we continue. Theorem 3.2 below established that in any case, the

vector $w = \bar{w}$ constitutes a subgradient of $f(\cdot)$ at the current point $\bar{\lambda}$. We

hence take a suitable step in the negative subgradient direction and project

back onto the feasible region Λ of Equation (3.63). This completes one iteration.

Before presenting Theorem 3.2, consider the following definition.

Definition 3.1

Let $F: \Lambda \to R$ be a convex function and let $\lambda \epsilon \Lambda \subset R^m$. Then $\xi \epsilon R^m$ is a

subgradient of $F(\cdot)$ at $\bar{\lambda}$ if

$$F(\lambda) \geq F(\bar{\lambda}) + \xi^t(\lambda - \bar{\lambda}) \text{ for each } \lambda \epsilon \Lambda.$$

Theorem 3.2

Let $\bar{\lambda}$ be a given point in Λ defined by (3.63) and let $\bar{w}$ be obtained from

Equations (3.57) through (3.62), with an arbitrary selection of a solution to

(3.59).

Then, $\bar{w}$ is a subgradient of $F(\cdot)$ at $\bar{\lambda}$, where $F: \Lambda \to R$ is defined in Equation (3.64).

Proof. Let y and $\bar{y}$ be obtained through Equation (3.57) from $\lambda \in \Lambda$ and $\bar{\lambda} \in \Lambda$ respectively. Hence,

$$F(\lambda) = \sum_{j=1}^{n} y_j^2 \text{ and } F(\bar{\lambda}) = \sum_{j=1}^{n} \bar{y}_j^2$$

Thus, from Definition 3.1, we need to show that

$$\sum_{h \in H} \sum_{i \in Q_h} \bar{w}_i^h (\lambda_i^h - \bar{\lambda}_i^h) \leq \sum_{j=1}^{n} y_j^2 - \sum_{j=1}^{n} \bar{y}_j^2 \tag{3.65}$$

Noting from Equations (3.52), (3.62) that $\sum_{h \in H} \sum_{i \in Q_h} \bar{w}_i^h \bar{\lambda}_i^h = 0$, we have,

$$\sum_{h \in H} \sum_{i \in Q_h} \bar{w}_i^h (\lambda_i^h - \bar{\lambda}_i^h) = \sum_{h \in H} \sum_{i \in Q_h} \bar{w}_i^h \lambda_i^h = \sum_{h \in H} \sum_{i \in Q_h} \sum_{j=1}^{n} \bar{u}_j^h \lambda_i^h [a_{ij}^h - \bar{y}_j]$$

$$= \sum_{h \in H} \sum_{j=1}^{n} \bar{u}_j^h \left(\sum_{i \in Q_h} \lambda_i^h a_{ij}^h \right) - \sum_{h \in H} \sum_{j=1}^{n} [\bar{u}_j^h \bar{y}_j \sum_{i \in Q_h} \lambda_i^h]$$

Using (3.48) and (3.50), this yields

$$\sum_{h \in H} \sum_{i \in Q_h} \bar{w}_i^h (\lambda_i^h - \bar{\lambda}_i^h) = \sum_{h \in H} \sum_{j=1}^{n} \bar{u}_j^h \left(\sum_{i \in Q_h} \lambda_i^h a_{ij}^h \right) - 2 \sum_{j=1}^{n} \bar{y}_j^2$$

Combining this with (3.65), we need to show that

$$\sum_{h \in H} \sum_{j=1}^{n} \bar{u}_j^h \left(\sum_{i \in Q_h} \lambda_i^h a_{ij}^h \right) \leq \sum_{j=1}^{n} y_j^2 + \sum_{j=1}^{n} \bar{y}_j^2 \tag{3.66}$$

But Equations (3.50), (3.55), (3.57) imply that

$$\sum_{h \in H} \sum_{j=1}^{n} \bar{u}_j^h \left(\sum_{i \in Q_h} \lambda_j^h a_{ij}^h \right) \leq \sum_{h \in H} \sum_{j=1}^{n} \bar{u}_j^h y_j = 2 \sum_{j=1}^{n} y_j \bar{y}_j$$

$$\leq 2 \| y \| \, \| \bar{y} \| \leq \| y \|^2 + \| \bar{y} \|^2$$

so that Equation (3.66) holds. This completes the proof.

Although, given $\bar{\lambda} \in \Lambda$, any solution to Equations (3.57) through (3.62) will yield a subgradient of $F(\cdot)$ at the current point $\bar{\lambda}$, we would like to generate, without expending much effort, a subgradient which is hopefully a direction of ascent. This would hence accelerate the cut generation process. Later in Section 3.4.1(b) we describe one such scheme to determine a suitable subgradient direction. For the present moment, let us assume that we have generated a subgradient $\bar{w}$ and have taken a suitable step size $\bar{\theta}$ in the direction $-\bar{w}$ as prescribed by the subgradient optimization scheme. Let

$$\bar{\bar{\lambda}} = \bar{\lambda} - \bar{\theta} \, \bar{w} \tag{3.67}$$

be the new point thus obtained. To complete the iteration, we must now project $\bar{\bar{\lambda}}$ onto Λ, that is, we must determine a new $\bar{\lambda}$ according to

$$\bar{\lambda}_{new} \equiv P_{\Lambda}(\bar{\bar{\lambda}}) = \text{minimum } \{\|\lambda - \bar{\bar{\lambda}}\| : \lambda \in \Lambda\} \tag{3.68}$$

The method of accomplishing this efficiently is presented in the next subsection.

3.4.1(c) Projection Scheme

For convenience, let us define the following linear manifold

$$M_h = \left\{ \lambda_i^h, \, i \in Q_h : \sum_{i \in Q_h} \lambda_i^h = 1 \right\}, \, h \in H \tag{3.69}$$

and let $\bar{M}_h$ be the intersection of M_h with the nonnegative orthant, that is,

$$\bar{M}_h = \{\lambda_i^h, \, i \in Q_h : \sum_{i \in Q_h} \lambda_i^h = 1, \, \lambda_i^h \geq 0, \, i \in Q_h\} \tag{3.70}$$

Note from Equation (3.63) that

$$\Lambda = \bar{M}_1 \times \ldots \times \bar{M}_{|H|} \tag{3.71}$$

Now, given $\bar{\bar{\lambda}}$, we want to project it onto Λ, that is, determine $\bar{\lambda}_{new}$ from Equation (3.68). Towards this end, for any vector $\alpha = (\alpha_i, i\epsilon I)$, where I is a suitable index set for the $|I|$ components of α, let $P(\alpha, I)$ denote the following problem

$$\underline{P(\alpha, I)}: \text{minimize } \{\frac{1}{2} \sum_{i\epsilon I} (\lambda_i - \alpha_i)^2 : \sum_{i\epsilon I} \lambda_i = 1, \lambda_i \geq 0, i\epsilon I\} \qquad (3.72)$$

Then to determine $\bar{\lambda}_{new}$, we need to find the solutions $(\bar{\lambda}_{new}^h)_i$, $i\epsilon Q_h$ as projections onto $\bar{M}_h$ of $\bar{\lambda}^h = (\bar{\bar{\lambda}}^h, i\epsilon Q_h)$ through each of the $|H|$ separable problems $P(\bar{\bar{\lambda}}^h, Q_h)$. Thus, henceforth in this section, we will consider only one such $h \in H$. Theorem 3.3 below is the basis of a finitely convergent iterative scheme to solve Problem $P(\bar{\bar{\lambda}}^h, Q_h)$.

Theorem 3.3

Consider the solution of Problem $P(\beta^k, I_k)$, where $\beta^k = (\beta^k, i\epsilon I_k)$, with $|I_k| \geq 1$. Define

$$\rho_k = (1 - \sum_{i\epsilon I_k} \beta_i^k) \Big/ |I_k| \qquad (3.73)$$

and let

$$\bar{\beta}^k = \beta^k + (\rho_k)1_k \qquad (3.74)$$

where 1_k denotes a vector of $|I_k|$ elements, each equal to unity. Further, define

$$I_{k+1} = \{i\epsilon I_k: \bar{\beta}_i^k > 0\} \qquad (3.75)$$

Finally, let β^{k+1} defined below be a subvector of $\bar{\beta}^k$,

$$\beta^{k+1} = (\beta_i^{k+1}, i\epsilon I_{k+1}) \qquad (3.76)$$

where, $\beta_i^{k+1} = \bar{\beta}_i^k$, $i\epsilon I_{k+1}$. Now suppose that $\hat{\beta}^{k+1}$ solves $P(\beta^{k+1}, I_{k+1})$.

(a) If $\bar{\beta}^k \geq 0$, then $\bar{\beta}^k$ solves $P(\beta^k, I_k)$

(b) If $\bar{\beta}^k \not\geq 0$, then β solves $P(\beta^k, I_k)$, where β has components given by

$$\beta_i = \begin{cases} \hat{\beta}_i^{k+1} & \text{, if } i\epsilon I_{k+1} \quad \text{ for each } i\epsilon I_k \\ \\ 0 & \text{otherwise} \end{cases} \tag{3.77}$$

<u>Proof.</u> For the sake of convenience, let $RP(\alpha,I)$ denote the problem obtained by relaxing the nonnegativity restrictions in $P(\alpha,I)$. That is, let

$$\underline{RP(\alpha,I)}: \quad \text{minimize } \{\frac{1}{2} \sum_{i\epsilon I} (\lambda_i - \alpha_i)^2 : \sum_{i\epsilon I} \lambda_i = 1\}$$

First of all, note from Equation (3.73), (3.74) that $\bar{\beta}^k$ solves $RP(\beta^k,I_k)$ since $\bar{\beta}^k$ is the projection of β^k onto the linear manifold

$$\{\lambda = (\lambda_i, i\epsilon I_k): \sum_{i\epsilon I_k} \lambda_i = 1\} \tag{3.78}$$

which is the feasible region of $RP(\beta^k,I_k)$. Thus, $\bar{\beta}^k \geq 0$ implies that $\bar{\beta}^k$ also solves $P(\beta^k,I_k)$. This proves part (a).

Next, suppose that $\bar{\beta}^k \not\geq 0$. Observe that β is feasible to $P(\beta^k,I_k)$ since from (3.77), we get $\beta \geq 0$ and $\sum_{i\epsilon I_k} \beta_i = \sum_{i\epsilon I_{k+1}} \hat{\beta}^{k+1} = 1$ as $\hat{\beta}^{k+1}$ solves $P(\beta^{k+1}, I_{k+1})$.

Now, consider any $\lambda = (\lambda_i, i\epsilon I_k)$ feasible to $P(\beta^k,I_k)$. Then, by the Pythagorem Theorem, since $\bar{\beta}^k$ is the projection of β^k onto (3.78), we get

$$\| \lambda - \beta^k \|^2 = \| \lambda - \bar{\beta}^k \|^2 + \| \bar{\beta}^k - \beta^k \|^2$$

Hence, the optimal solution to $P(\bar{\beta}^k,I_k)$ is also optimal to $P(\beta^k,I_k)$. Now, suppose that we can show that the optimal solution to Problem $P(\bar{\beta}^k,I_k)$ must satisfy

$$\lambda_i = 0 \text{ for } i\notin I_{k+1} \tag{3.79}$$

Then, noting (3.76), (3.77), and using the hypothesis that $\hat{\beta}^{k+1}$ solves $P(\beta^{k+1},I_{k+1})$, we will have established part (b). Hence, let us prove that (3.79) must hold. Towards this end, consider the Karush–Kuhn–Tucker equations for Problem $P(\bar{\beta}^k, I_k)$ with t and w_i, $i\varepsilon I_k$ as the appropriate langrangian multipliers.

$$\sum_{i\varepsilon I_k} \lambda_i = 1, \; \lambda_i \geq 0 \text{ for each } i\varepsilon I_k \tag{3.80}$$

$$(\lambda_i - \bar{\beta}_i^k) + t - w_i = 0 \text{ and } w_i \geq 0 \text{ for each } i\varepsilon I_k \tag{3.81}$$

$$\lambda_i w_i = 0 \text{ for each } i\varepsilon I_k \tag{3.82}$$

Now, since $\sum\limits_{i\varepsilon I_k} \bar{\beta}_i^k = 1$, we get from (3.80), (3.81) that

$$t = \sum_{i\varepsilon I_k} w_i \bigg/ |I_k| \geq 0 \tag{3.83}$$

But from (3.81), (3.82) we get for each $i\varepsilon I_k$,

$$0 = w_i \lambda_i = \lambda_i(\lambda_i + t - \bar{\beta}_i^k)$$

which implies that for each $i\varepsilon I_k$, we must have,

either $\lambda_i = 0$, whence from (3.81), $w_i = t - \bar{\beta}_i^k$ must be nonnegative

or $\lambda_i = \bar{\beta}_i^k - t$, whence from (3.81), $w_i = 0$.

In either case above, noting (3.80), if $\bar{\beta}_i^k \leq 0$, that is, if $i\notin I_{k+1}$, we must have $\lambda_i = 0$. This completes the proof.

Using Theorem 3.3, one may easily validate the following procedure for finding $\bar{\lambda}_{new}^h$ of Equation (3.68), given $\bar{\bar{\lambda}}^h$. This procedure has to be repeated separately for each $h\varepsilon H$.

Initialization

Set $k=0$, $\beta^0 = \bar{\bar{\lambda}}^h$, $I_0 = Q_h$. Go to Step 1.

Step 1

Given β^k, I_k, determine ρ_k and $\bar{\beta}^k$ from (3.73), (3.74). If $\beta^k \geq 0$, then terminate with $\bar{\lambda}_{new}^h$ having components given by

$$(\bar{\lambda}^h_{new})_i = \begin{cases} \bar{\beta}^k_i & \text{if } i\varepsilon I_k \\ \\ 0 & \text{otherwise} \end{cases}$$

Otherwise, proceed to Step 2.

Step 2

Define I_{k+1}, β^{k+1} as in Equations (3.75), (3.76), increment k by one and return to Step 1.

Note that this procedure is finitely convergent as it results in a strictly decreasing, finite sequence $|I_k|$ satisfying $|I_k| \geq 1$ for each k, since $\sum_{i\varepsilon I_k} \bar{\beta}^k = 1$ for each k.

Example

Suppose we want to project $\bar{\bar{\lambda}}^h = (-2,3,1,2)$ on to $\Lambda \subset R^4$. Then the above procedure yields the following results.

Initialization

k=0, $\beta^0 = (-2,3,1,2)$, $I_0 = \{1,2,3,4\}$.

Step 1

$\rho_0 = -3/4$, $\bar{\beta}^0 = (-\frac{11}{4}, \frac{9}{4}, \frac{1}{4}, \frac{5}{4})$

Step 2

k=1, $I_1 = \{2,3,4\}$, $\beta^1 = (\frac{9}{4}, \frac{1}{4}, \frac{5}{4})$

Step 1

$\rho_1 = -\frac{11}{12}$, $\bar{\beta}^1 = (\frac{4}{3}, -\frac{2}{3}, \frac{1}{3})$

Step 2

k=2, $I_2 = \{2,4\}$, $\beta^2 = (\frac{4}{3}, \frac{1}{3})$

Step 1

$\rho_2 = -\frac{1}{3}$, $\bar{\beta}^2 = (1,0) \geq 0$

Thus, $\bar{\lambda}^h_{new} = (0,1,0,0)$.

3.4.1(d) A Second Sufficient Condition for Termination

As indicated earlier in Section 3.4.1(b), we will now derive a second sufficient condition on $\bar{w}$ for $\bar{\lambda}$ to solve PD_2. For this purpose, consider the following lemma.

Lemma 3.4

Let $\bar{\lambda} \in \Lambda$ be given and suppose we obtain $\bar{w}$ using Equations (3.57) through (3.62). Let $\hat{w}$ solve the problem.

$$\underline{PR_h}: \quad \text{minimize} \left\{ \frac{1}{2} \sum_{i \in Q_h} (\bar{w}_i^h - w_i^h)^2 : \sum_{i \in Q_h} w_i^h = 0, \; w_i^h \leq 0 \text{ for } i \in J_h \right\}$$

$$\text{for each } h \in H$$

where,

$$J_h = \{i \in Q_h : \bar{\lambda}_i^h = 0\}, \; h \in H \tag{3.84}$$

Then, if $\hat{w} = 0$, $\bar{\lambda}$ solves Problems PD_2.

Proof. Since $\hat{w} = 0$ solves PR_h, $h \in H$, we have for each $h \in H$,

$$\sum_{i \in Q_h} (\bar{w}_i^h)^2 \leq \sum_{i \in Q_h} (\bar{w}_i^h - w_i^h)^2 \tag{3.85}$$

for all w_i^h, $i \in Q_h$ satisfying $\sum_{i \in Q_h} w_i^h = 0$, $w_i^h \leq 0$ for $i \in J_h$. Given any $\lambda \in \Lambda$ and given any $\mu > 0$ define

$$w_i^h = (\bar{\lambda}_i^h - \lambda_i^h) \big/ \mu, \; i \in Q_h, \; h \in H \tag{3.86}$$

Then, $\sum_{i \in Q_h} w_i^h = 0$ for each $h \in H$ and since $\bar{\lambda}_i^h = 0$ for $i \in J_h$, $h \in H$, we get $w_i^h \leq 0$ for $i \in J_h$, $h \in H$. Thus, for any $\lambda \in \Lambda$, but substituting (3.86) into (3.85), we have,

$$\mu^2 \sum_{i \in Q_h} (\bar{w}_i^h)^2 \leq \sum_{i \in Q_h} (\lambda_i^h - \bar{\lambda}_i^h + \mu\bar{w}_i^h)^2 \text{ for each } h \in H \tag{3.87}$$

But equation (3.87) implies that for each $h \in H$, $\lambda^h = \bar{\lambda}^h$ solves the problem

$$\text{minimize} \left\{ \sum_{i \in Q_h} \{[\lambda_i^h - (\bar{\lambda}_i^h - \bar{w}_i^h)]\}^2 : \sum_{i \in Q_h} \lambda_i^h = 1, \; \lambda_i^h \geq 0, \; i \in Q_h \right\} \text{for each } h \in H$$

In other words, the projection $P_\Lambda(\bar{\lambda} - \bar{w}\mu)$ of $(\bar{\lambda} - \bar{w}\mu)$ onto Λ is equal to $\bar{\lambda}$ for any $\mu > 0$. From the theory of subgradient optimization, since $\bar{w}$ is a subgradient of $F(\cdot)$ at $\bar{\lambda}$, then $\bar{\lambda}$ solves PD_2. This completes the proof.

Note that Lemma 3.4 above states that if the "closest" feasible direction $-w$ to $-\bar{w}$ is a zero vector, then $\bar{\lambda}$ solves PD_2. Based on this result, we derive through Lemma 3.5 below a second sufficient condition for $\bar{\lambda}$ to solve PD_2.

Lemma 3.5

Suppose w=0 solves Problems PR_h, $h \in H$ as in Lemma 3.4. Then for each $h \in H$, we must have

(a) $\bar{w}_i^h = t_h$, a constant, for each $i \notin J_h$

(3.88)

(b) $\bar{w}_i^h \leq t_h$ for each $i \in J_h$

where J_h is given by Equation (3.84).

Proof. Let us write out the Karush-Kuhn-Tucker conditions for Problem PR_h, for any $h \in H$. We obtain

$(w_i^h - \bar{w}_i^h) + t_h = 0$ for $i \notin J_h$

$(w_i^h - \bar{w}_i^h) + t_h - u_i^h = 0$ for $i \in J_h$

$u_i^h \geq 0, \; i \in J_h, \; u_i^h w_i^h = 0 \; i \in J_h$

$\sum_{i \in Q_h} w_i^h = 0, \; w_i^h \geq 0$ for $i \in J_h$, t_h unrestricted

If w=0, solves PR_h, $h \in H$, then since PR_h has a convex objective function and linear constraints, then there must exist a solution to

$$\bar{w}_i^h = t_h \text{ for each } i \notin J_h$$

and

$$u_i^h = (t_h - \bar{w}_i^h) \geq 0 \text{ for each } i \varepsilon J_h.$$

This completes the proof.

Thus Equation (3.88) gives us another sufficient conditon for $\bar{\lambda}$ to solve PD_2. We illustrate the use of this condition through an example in Section 3.4.1(g).

3.4.1(e) Schema of an Algorithm to Solve Problem PD_2

The procedure developed above is depicted schematically in Figure 3.3. In block 1 an arbitrary, or preferably a good heuristic solution, $\bar{\lambda} \varepsilon \Lambda$ is sought. For example, one may use $\bar{\lambda}_i^h = 1 \big/ |Q_h|$ for each $i \varepsilon Q_h$, for $h \varepsilon H$. For blocks 4 and 6, we recommend the standard procedural steps adopted for the subgradient optimization scheme.

3.4.1(f) Derivation of a Good Subgradient Direction

In our discussion in Section 3.4.1(a), we saw that given a $\lambda \varepsilon \Lambda$ of Equation (3.63), we were able to uniquely determine $\bar{y}_j$, j=1,...,n through Equation (3.57). Thereafter, once we fixed values $\bar{u}_j^h$ for u_j^h, j=1,...,n, $h \varepsilon H$ satisfying Equation (3.59), we were able to uniquely determine values for the other variables in the Karush-Kuhn-Tucker System using Equations (3.61), (3.62). Moreover, the only choice in determining $\bar{u}_j^h$, j=1,...,n, $h \varepsilon H$ arose in case $|H_j| \geq 2$ for some $j \varepsilon \{1,...,n\}$ in Equation (3.60). We also established that no matter what feasible values we selected for u_j^h, $j \varepsilon \{1,...,n\}$, $h \varepsilon H$, the corresponding vector w obtained was a subgradient direction. In order to select the best such subgradient direction, we are interested in finding a vector $\bar{w}$ which has the smallest euclidean norm among all possible vectors corresponding to the given solution $\bar{\lambda} \varepsilon \Lambda$. However, this problem is not easy to solve. Moreover, since this step will merely be a subroutine at each iteration of the proposed scheme to solve PD_2, we will present a heuristic approach to this problem.

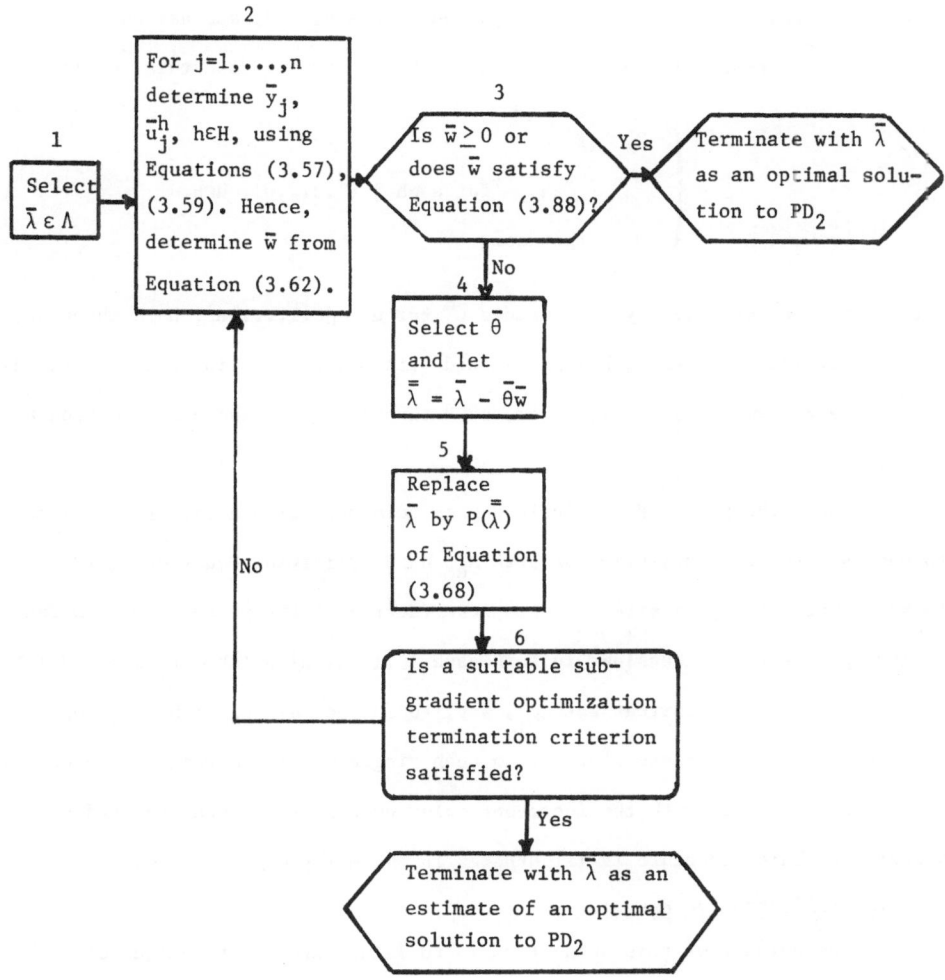

Figure 3.3. Schema of an Algorithm for Problem PD_2

Towards this end, let us define for convenience, mutually exclusive but not uniquely determined sets N_h, $h \in H$ as follows

$$N_h \subset \{j \in \{1,\ldots,n\}: h \in H_j \text{ of Equation (3.58)}\} \qquad (3.89)$$

$$N_i \cap N_j = \{\phi\} \text{ for any } i, j \in H \text{ and } \bigcup_{h \in H} N_h = \{j \in \{1,\ldots,n\}: \bar{y}_j > 0\} \qquad (3.90)$$

In other words, we take each $j\varepsilon\{1,\ldots,n\}$ which has $\bar{y}_j > 0$, and assign it to some $h\varepsilon H_j$, that is, assign it to a set N_h, where $h\varepsilon H_j$. Having done this, we let

$$\bar{u}_j^h = \begin{cases} 2\bar{y} & \text{if } j\varepsilon N_h \\ \\ 0 & \text{otherwise} \end{cases} \qquad \text{for each } j\varepsilon\{1,\ldots,n\}, \; h\varepsilon H. \qquad (3.91)$$

Note that Equation (3.91) yields values $\bar{u}_j^h$ for u_j^h, $j\varepsilon\{1,\ldots,n\}$, $h \varepsilon H$ which are feasible to (3.59). Hence, having defined sets N_h, $h \varepsilon H$ as in Equations (3.89), (3.90), we determine $\bar{u}_j^h$, $j\varepsilon\{1,\ldots,n\}$, $h \varepsilon H$ through (3.91) and hence $\bar{w}$ through (3.62).

Thus, the proposed heuristic scheme commences with a vector w obtained through an arbitrary selection of sets N_h, $h \varepsilon H$ satisfying Equations (3.89), (3.90). Thereafter, we attempt to improve (decrease) the value of $w^t w$ in the following manner. We consider in turn each $j\varepsilon\{1,\ldots,n\}$ which satisfies $|H_j| \geq 2$ and move it from its current set N_{h_j}, say, to another set N_h with $h\varepsilon H_j$, $h\neq h_j$, if this results in a decrease $w^t w$. If no such single movements result in a decrease in $w^t w$, we terminate with the incumbent solution w as the sought subgradient direction. This procedure is illustrated in the example given below.

3.4.1(g) Illustrative Example

The purpose of this subsection is to illustrate the technique of the foregoing section for determining a good subgradient direction as well as the termination criterion of Section 3.4.1(d).

Thus, let $H = \{1,2\}$, n=3, $|Q_1| = |Q_2| = 3$ and consider the constraint sets

$$S_1 = \left\{ x: \begin{array}{r} 2x_1 - 3x_2 + x_3 \geq 1 \\ -x_1 + 2x_2 + 3x_3 \geq 1 \\ 3x_1 - x_2 - x_3 \geq 1 \\ x_1, x_2, x_3 \geq 0 \end{array} \right\} \quad \text{and } S_2 = \left\{ x: \begin{array}{r} 3x_1 - x_2 - x_3 \geq 1 \\ 2x_1 + x_2 - 2x_3 \geq 1 \\ -x_1 + 3x_2 + 3x_3 \geq 1 \\ x_1, x_2, x_3 \geq 0 \end{array} \right\}$$

Further, suppose we are currently located at a point $\bar{\lambda}$ with

$$\bar{\lambda}_1^1 = 0, \ \bar{\lambda}_2^1 = 5/12, \ \bar{\lambda}_3^1 = 7/12; \ \bar{\lambda}_1^2 = 7/12, \ \bar{\lambda}_2^2 = 0, \ \bar{\lambda}_3^2 = 5/12.$$

Then the associated surrogate constraints are

$$\frac{4}{3} x_1 + \frac{1}{4} x_2 + \frac{2}{3} x_3 \geq 1 \quad \text{for h=1}$$

and $\hspace{10cm}$ (3.92)

$$\frac{4}{3} x_1 + \frac{2}{3} x_2 + \frac{2}{3} x_3 \geq 1 \quad \text{for h=2}$$

Using Equations (3.57), (3.60), we find

$$\bar{y}_1 = \frac{4}{3} \text{ with } H_1 = \{1,2\}, \ \bar{y}_2 = \frac{2}{3} \text{ with } H_2 = \{2\} \text{ and } \bar{y}_3 = \frac{2}{3} \text{ with } H_3 = \{1, 2\}.$$

Note that the possible combinations of N_1 and N_2 are as follows:

$$\text{(i) } N_1 = \{1\}, \ N_2 = \{2,3\},$$

$$\text{(ii) } N_1 = \{\phi\}, \ N_2 = \{1,2,3\},$$

$$\text{(iii) } N_1 = \{1,3\}, \ N_2 = \{2\}, \text{ and}$$

$$\text{(iv) } N_1 = \{3\}, \ N_2 = \{1,2\}.$$

A total enumeration of the values of u obtained for these sets through (3.91) and the corresponding values for w are shown below.

N_1	N_2	u_1^1	u_2^1	u_3^1	u_1^2	u_2^2	u_3^2	w_1^1	w_2^1	w_3^1	w_1^2	w_2^2	w_3^2	$w^t w$
		$u_j^h, j\varepsilon\{1,\ldots,n\}$						$w_i^h, i\varepsilon Q_h, h\varepsilon H$						
{1}	{2,3}	8/3	0	0	0	4/3	4/3	16/9	-56/9	40/9	-40/9	-28/9	56/9	129.78
{φ}	{1,2,3}	0	0	0	8/3	4/3	4/3	0	0	0	0	- 4/3	0	1.78
{1,3}	{2}	8/3	0	4/3	0	4/3	0	20/9	-28/9	20/9	-20/9	4/9	28/9	34.37
{3}	{1,2}	0	0	4/3	8/3	4/3	0	-4/9	28/9	-20/9	20/9	20/9	-28/9	34.37

Thus, according to the proposed scheme, if we commence with $N_1 = \{1\}$, $N_2 = \{2,3\}$, then picking j=1 which has $|H_j| = 2$, we can move j=1 into N_2 since $2 \in H_1$. This leads to an improvement. As one can see from above, no further improvement is possible. In fact, the best solution shown above is accessible by the proposed scheme by all except the third case which is a "local optimal".

We now illustrate the sufficient termination condition of Section 3.4.1(d). The vector $\bar{w}$ obtained above is $(0,0,0 \overset{h=1}{|} 0,\frac{-4}{3},0)$. Further the vector $\bar{\lambda}$ is $(0, \frac{5}{12}, \frac{7}{12} \overset{h=1}{|} \frac{7}{12}, \overset{h=2}{0}, \frac{-5}{12})$. Thus, even though $\bar{w} \not\geq 0$, we see that the conditions (3.88) of Lemma 3.5 are satisfied for each $h \in H = \{1,2\}$ and thus the given $\bar{\lambda}$ solves PD$_2$.

The disjunctive cut (3.19) derived with this optimal solution $\bar{\lambda}$ is obtained through (3.92) as

$$\frac{4}{3} x_1 + \frac{2}{3} x_2 + \frac{2}{3} x_3 \geq 1 \tag{3.93}$$

3.4.2 Maximizing the Rectilinear Distance Between the Origin and the Disjunctive Cut

In this section, we will briefly consider the case where one desires to use rectilinear instead of euclidean distances. Extending the developments of Sections 3.2, 3.3 and 3.4.1, one may easily see that the relevant problem is

$$\text{minimize}\{ \ \underset{j\epsilon\{1,\ldots,n\}}{\text{maximum}} \ y_j : \text{constraints (3.47), (3.48), (3.49) are satisfied}\}.$$

The reason why we consider this formulation is its intuitive appeal. To see this, note that the above problem is separable in $h \epsilon H$ and may be rewritten as

$$\text{PD}_1: \ \text{minimize}\{\xi^h : \xi^h \geq \sum_{i\epsilon Q_h} \lambda_i^h a_{ij}^h \ \text{for each } j=1,\ldots,n, \ \sum_{i\epsilon Q_h} \lambda_i^h = 1, \ \lambda_i^h \geq 0$$

$$\text{for each } i\epsilon Q_h, \ \xi^h \geq 0\} \quad \text{for each } h \epsilon H.$$

Thus, for each $h \epsilon H$, PD_1 seeks λ_i^h, $i\epsilon Q_h$ such that the largest of the surrogate constraint coefficients is minimized. Once such surrogate constraints are obtained, the disjunctive cut (3.19) may be derived using the principles of Section 3.3.

As far as the solution of Problem PD_1 is concerned, we merely remark that one may either solve it as a linear program or rewrite it as the minimization of a piecewise linear convex function subject to linear constraints and use a subgradient optimization technique. We further note that the structure of Problem PD_1 may render it more amenable to the latter solution technique.

3.5 Other Criteria for Obtaining Deep Cuts

In this section, we will briefly deal with some other plausible criteria which one may adopt. Since DC1 is a specia case of DC2, we will treat only the latter case. Further, we will consider the original disjunction DC2, that is, we will not require $b_i^h > 0$ for each $i \epsilon Q_h$, h H. Note that the basic disjunctive cut for DC2 is given by Equation (2.10) rewritten below for convenience

$$\sum_{j=1}^{n} [\max_{h\epsilon H} \{ \sum_{i\epsilon Q_h} \lambda_i^h a_{ij}^h \}] \ x_j \geq \min_{h\epsilon H} \{ \sum_{i\epsilon Q_h} \lambda_i^h b_i^h\} \tag{3.94}$$

Now, a criterion which may be suggested would be to maximize the surplus with respect to the origin. This would mean that the right-hand-side of (3.94) should be made as large as possible, with, of course, some restriction on the overall magnitude of the minimand such as

$$\sum_{i \in Q_h} \sum_{h \in H} \lambda_i^h b_i^h = \hat{h}, \text{ say} \qquad (3.95)$$

where,

$$\hat{h} = |H|.$$

One may easily verify that this implies it is optimal to select

$$\sum_{i \in Q_h} \lambda_i^h b_i^h = 1 \text{ for each } h \in H. \qquad (3.96)$$

Other than the restriction (3.96), we are still free to select nonnegative values for λ_i^h, $i \in Q_h$, $h \in H$. Since the resultant cut (3.94) should at least support S (Equation 3.11)), we may simply select a set of positive coefficients λ_j, $j \in N$ and solve the linear program.

LP: minimize $\displaystyle \sum_{j \in N} \delta_j y_j$

subject to $\displaystyle \sum_{i \in Q_h} \lambda_i^h b_i^h = 1$ for each $h \in H$

$\displaystyle \sum_{i \in Q_h} \lambda_i^h a_{ij}^h \leq y_j$ for each $h \in H$, for $j = 1, \ldots, n$

$\lambda_i^h \geq 0$, $i \in Q_h$, $h \in H$

y_j, $j \in N$ are unrestricted in sign.

Essentially, the constraints of Problem LP conform with those of Problem PD_2 (Equations (3.47), (3.48), (3.49)). Choosing different values for $\delta_j > 0$, $j \in N$ in the objective function of LP would yield different cuts (through the parameters λ_i^h, $i \in Q_h$, $h \in H$), all of which would be nondominated supports of S, including facets of S.

3.6 Some Standard Choices of Surrogate Constraint Multipliers

We now present two standard procedures for selecting values for the

parameters λ_i^h, $i \in Q_h$, $h \in H$ for the disjunction DC2. Although not strongly motivated,

these solution procedures have intuitive appeal.

The first of these methods prescribes that the constraint sets S_h, $h \in H$

be first represented in the form given in Equation (3.36), and then one may select

$\lambda_i^h = 1 \Big/ |Q_h|$ for each $i \in Q_h$, for $h \in H$.

As a second method, one may identify for each $h \in H$, a constraint which has

the largest number of minimal column elements. That is, for each $h \in H$, we compute

minimum $\{a_{ij}^h : i \in Q_h\}$ for each $j \in N$ and identify the constraint $\hat{i} \in Q_h$ which contains

the maximum number of these $|N|$ minimal coefficients. On the other hand, we may

let $\hat{i}$ be the constraint with the most number of negative coefficients for each

$h \in H$. We then set $\lambda_{\hat{i}}^h = 1$ and $\lambda_i^h = 0$ for $i \in Q_h$, $i \neq \hat{i}$, for each $h \in H$.

Before illustrating the above two methods, we draw the readers' attention

to an obvious, though pertinent, fact. Suppose we are given constraint sets S_h,

$h \in H$ of Equation (3.12) as in DC1 and the disjunction states that at least k of

these constraints must be satisfied, where $k < |H|$. Then, by grouping the $|H|$

sets k at a time, we may equivalently represent this disjunction as DC2, stating

that at least one of the resulting $\binom{|H|}{k}$ sets must be satisfied. A disjunctive

cut may now be derived based on the statement DC2. There is, however, an

alternative approach. Note that we may choose to delete any (k-1) of the $|H|$

constraint sets and then assert that at least one of the remaining sets must be

satisfied. This would then represent the given disjunction as DC1. Of course,

for k=1, both the alternatives are identical. For smaller values of k, the latter

technique is likely to be superior to the former technique since by deleting the

rows which contain the largest number of column-maxima, one can usually do

better than say, by averaging coefficients. On the other hand, for k close to

$|H|$, the former is likely to be better since by taking the average, say, of

several k numbers of arbitrary sign would tend to produce smaller cut coefficients.

These are simply general rules of thumb and clearly one may produce examples which indicate the contrary. We now illustrate the two methods proposed above for selecting values for λ_i^h, $i \varepsilon Q_h$, $h \varepsilon H$.

Example

Consider the example of Section 3.4.1(g). The first method discussed above yields $\lambda_i^h = 1/3$ for each $i \varepsilon Q_h$, $h \varepsilon H$. This gives surrogate constraints

$$\frac{4}{3} x_1 - \frac{2}{3} x_2 + x_3 \geq 1 \text{ for } h = 1 \text{ and } \frac{4}{3} x_1 + x_2 \geq 1 \text{ for } h = 2.$$

Hence, the disjunctive cut is

$$\frac{4}{3} x_1 + x_2 + x_3 \geq 1 \tag{3.97}$$

This cut is uniformly dominated by the cut (3.93) derived through Problem PD2.

The second method discussed above suggests that we should use $\lambda_3^1 = \lambda_1^2 = 1$ and $\lambda_i^h = 0$ otherwise for h=1,2, and i=1,2,3. This yields the cut

$$3x_1 - x_2 - x_3 \geq 1 \tag{3.98}$$

Neither (3.98) nor (3.93) uniformly dominate the other. However, the values of the euclidean and the rectilinear distance criterion for each of these cuts (3.93) and (3.98) are respectively 0.6123, 0.75 and 0.111, 0.333.

3.7 Note and References

The question of how to specify the cut parameters λ_i^h has been addressed before in the general context of cutting plane theory. However, Balas [7] and Glover [19] have addressed the question of finding these parameters in the context of disjunctive programs. The parameters defined in equation (3.15) was used by Balas [6]. The cut defined by (3.19) and (3.20) was motivated by a similar result due to Glover [18] in the context of convexity cuts. This chapter contains several results in the context of disjunctive programming due to appear in [ ].

Chapter IV

EFFECT OF DISJUNCTIVE STATEMENT FORMULATION ON DEPTH OF
CUT AND POLYHEDRAL ANNEXATION TECHNIQUES

4.1 Introduction

In this chapter, we wish to emphasize two important salient features of
disjunctive programming methods. Both these features basically relate to the
issue of depth of cut. More specifically, we will first illustrate that one
can derive cuts differing in depth through different formulations of a given
disjunctive statement. Secondly, we will exhibit some connections between disjunctive programming technqiues and known polyhedral annexation methods. Based
on the latter exposition, as well as on some further development, we will exhibit
how one may strike a reasonable tradeoff between the effort involved to generate
a cut and its depth.

The organization of this chapter is as follows. First, we illustrate
the above tradeoff involved through a numerical example. Thereafter, we make
some general remarks and in particular we relate these ideas to two specific
cases, namely, the generalized lattice point problem and the linear complementarity problem. Next, as in Chapter III, we consider two situations – one in
which each set S_h, $h \in H$ (Equation (1.1)) contains exactly one constraint and a
second case in which each set S_h, $h \in H$ may contain more than one constraint.
Using the first case, we establish connections between disjunctive programming
methods and polyhedral annexation techniques. Following this, we demonstrate
two schemes by which improved disjunctive cuts may be derived through suitable
disjunction formulations. Finally, we present extensions of these developments
to the second case.

4.2 Illustration of the Tradeoff Between Effort for Cut Generation and the Depth
     of Cut

Consider the problem

$$\text{maximize} \quad 2x_1 + 3x_2$$

subject to $x_1 + x_2 \leq 10$ or $x_1 + x_2 + s_1 = 10$

$x_1 \qquad \leq 8$ or $x_1 \qquad + s_2 = 8$

$x_2 \leq 5$ or $\qquad x_2 + s_3 = 5$

$x_1, x_2 \geq 0, \qquad s_1, s_2, s_3 \geq 0$

Further, suppose that the following disjunctive statement must hold:

{Either x_1 or x_2 must equal zero, i.e., $x_1 x_2 = 0$}

Relaxing the disjunctive statement and solving the corresponding linear program, we obtain the solution depicted in Figure 4.1.

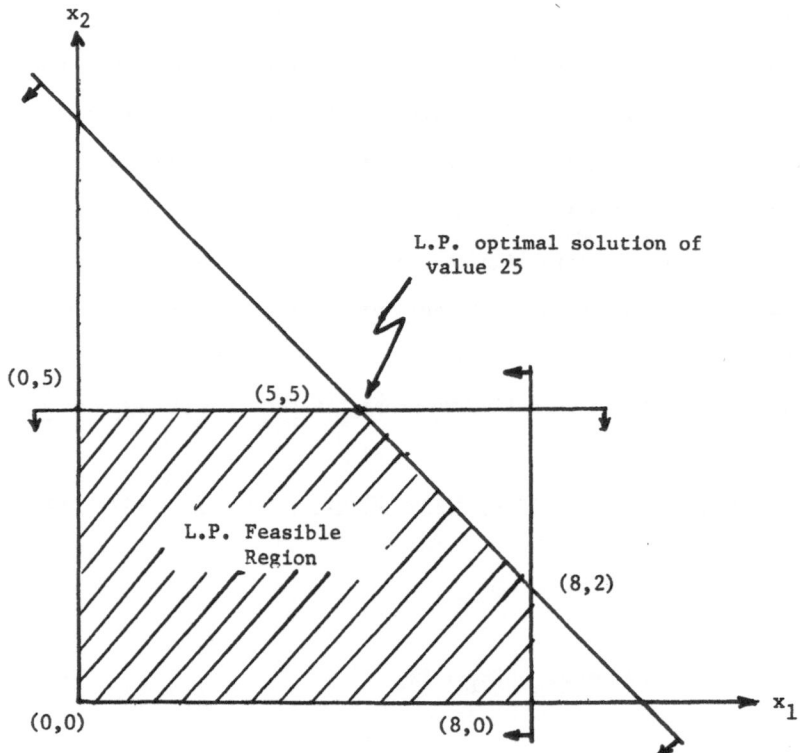

Figure 4.1.   LP Solution with the Disjunction Relaxed

The simplex tableau corresponding to this optimal linear programming solution, which incidentally is feasible to the disjunction, is given below.

		Non-basic variables		RHS
		s_1	s_3	
objective function		2	1	25
Basic Variables	x_1	1	-1	5
	s_2	-1	1	3
	x_2	0	1	5

From this tableau, we can represent x_1 and x_2 in terms of the non-basic variables as

$$x_1 = 5 - s_1 + s_3$$

$$x_2 = 5 - s_3 \tag{4.1}$$

Hence, the disjunctive statement $x_1 x_2 = 0$ may be restated as follows. At least one of the constraints $x_1 \leq 0$ or $x_2 \leq 0$ must be satisfied along with nonnegativity restrictions. In terms of the current nonbasic variables, this may be restated as implying that at least one of the following constraint sets must be satisfied

$$\{(s_1, s_3): s_1 - s_3 \geq 5, \ s_1, s_3 \geq 0\}$$

$$\{(s_1, s_3): \quad s_3 \geq 5, \ s_1, s_3 \geq 0\}. \tag{4.2}$$

Now, from our analysis in Chapter III, we know that the best cut which one may derive from this disjunction is

$$[\max\{\tfrac{1}{5}, 0\}]s_1 + [\max\{-\tfrac{1}{5}, \tfrac{1}{5}\}]s_3 \geq 1$$

$$\text{i.e.} \quad s_1 + s_3 \geq 5 \qquad\qquad (4.3)$$

This cut may now be appended to the above tableau and the optimization procedure continued. The reader may note that, we can use Equation (4.1) to re-write the cut (4.3) as

$$x_1 + 2x_2 \leq 10 \qquad\qquad (4.4)$$

Now, let us get a geometric interpretation as to why (4.3) is indeed the deepest cut. Note that in specifying the disjunction (4.2), we have neglected nonnegativity on s_2 in the tableau representing the current point. Effectively, we neglected the constraint $x_1 \leq 8$ and used only "local" information. As a result, we implied that the feasible region of the problem is $\{x_1 = 0, 0 \leq x_2 \leq 5\} \cup \{x_2 = 0, 0 \leq x_1 \leq 10\}$. The convex hull of this region is S' and is depicted in Figure 4.2 below. One may observe from S' that the best corresponding cut is precisely cut (4.4)

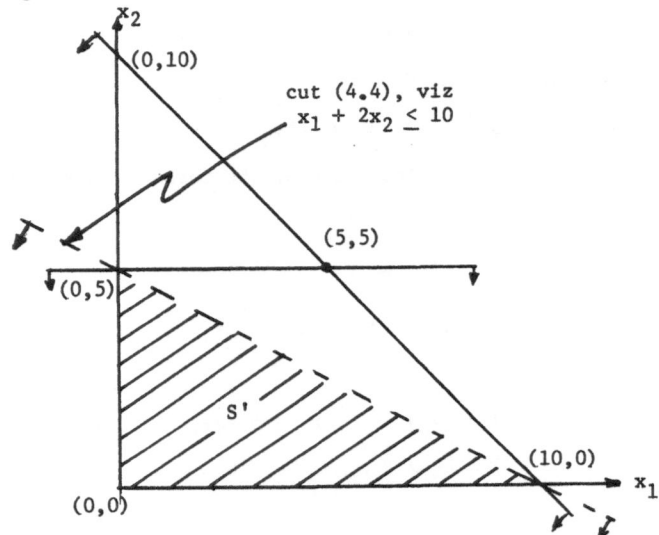

Figure 4.2. Deepest Cut for the Formulation (4.2) of the Given Disjunction

Now, let us specify the disjunction using additional information. We know for the above example problem that we must have either $x_1 = 0$ whence, $0 \leq x_2 \leq 5$ or we must have $x_2 = 0$ whence, $0 \leq x_1 \leq 8$. This feasible region to the example problem is shown darkened in Figure 4.3. Now, any valid cut should not delete any points in this feasible region. Since the half-space feasible to a cut is a closed convex set, the cut must not delete any point in the convex hull S of this feasible region shown thatched in Figure 4.3. Hence, a deep cut can at best support S, and the best cut in the present context is clearly $5x_1 + 8x_2 \leq 40$. This cut as well as the cut (4.4) is shown in Figure 4.3.

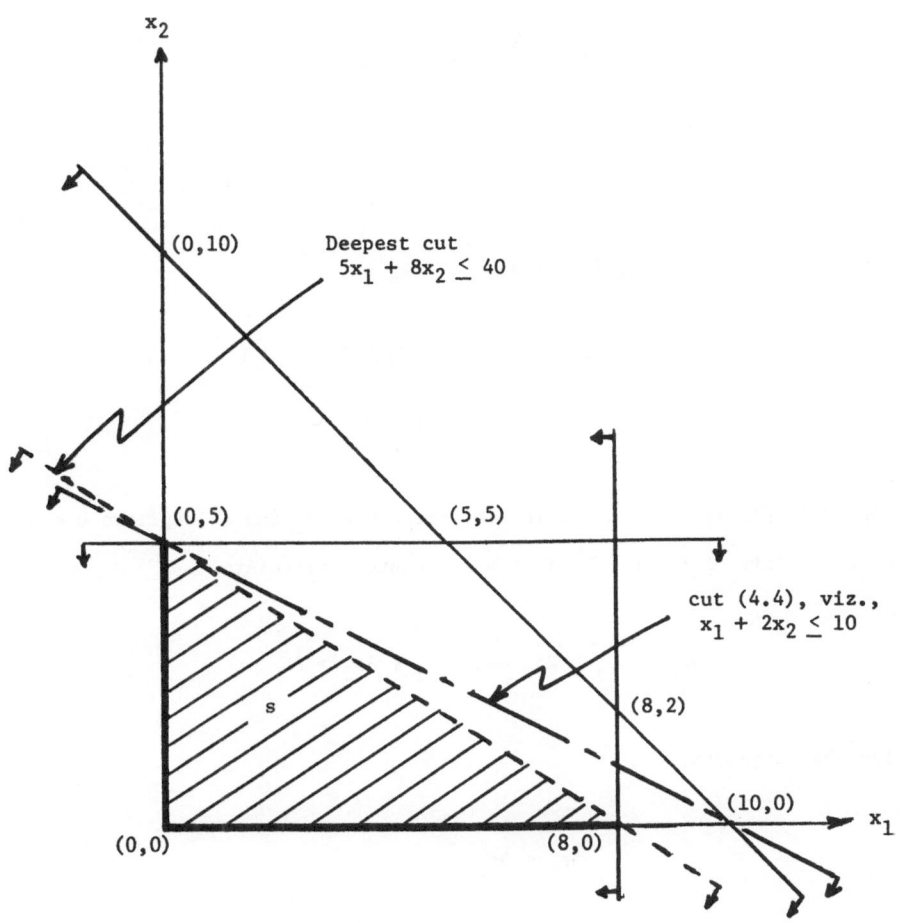

Figure 4.3. Deepest Cut

We will now see how the deepest cut $5x_1 + 8x_2 \leq 40$ can be derived algebraically. Obviously, to obtain this cut we must consider all the constraints of the original problem. In other words, the nonnegativity restrictions on s_2 must be included as well. Since

$$s_2 = 3 + s_1 - s_3 \qquad\qquad (4.5)$$

in the tableau representing the current point, the given disjunctive statement implies that at least one of the following constraint sets must be satisfied.

$$
\begin{aligned}
S_1\colon \{(s_1, s_3)\colon \quad & s_1 - s_3 \geq 5 \\
& s_1 - s_3 \geq -3 \\
& s_1, s_3 \geq 0\} \\
S_2\colon \{(s_1, s_3)\colon \quad & s_3 \geq 5 \\
& s_1 - s_3 \geq -3 \\
& s_1, s_3 \geq 0\}
\end{aligned}
\qquad (4.6)
$$

Using multipliers 5 and 0 for the constraints of S_1 and multipliers 8 and 5 for the constraints of S_2, we obtain the surrogate constraints

$$5s_1 - 5s_3 \geq 25 \quad \text{and} \quad 5s_1 + 3s_3 \geq 25$$

This yields the disjunctive cut

$$5s_1 + 3s_3 \geq 25 \qquad\qquad (4.7)$$

or, using Equations (4.1), this may be rewritten as

$$5x_1 + 8x_2 \leq 40 \qquad\qquad (4.8)$$

As depicted in Figure 4.3, the cut (4.8) is the deepest possible for the given disjunction. Later in Sections 4.4 and 4.5, we will show how (4.7) or (4.8) may be derived conveniently through (4.6).

So far we have illustrated the cuts in the (x_1,x_2) space. In the space defined by the nonbasic variables (s_1,s_3), the cuts are as given by (4.3) and (4.7). These cuts are illustrated in Figures 4.4(a) and 4.4(b) respectively.

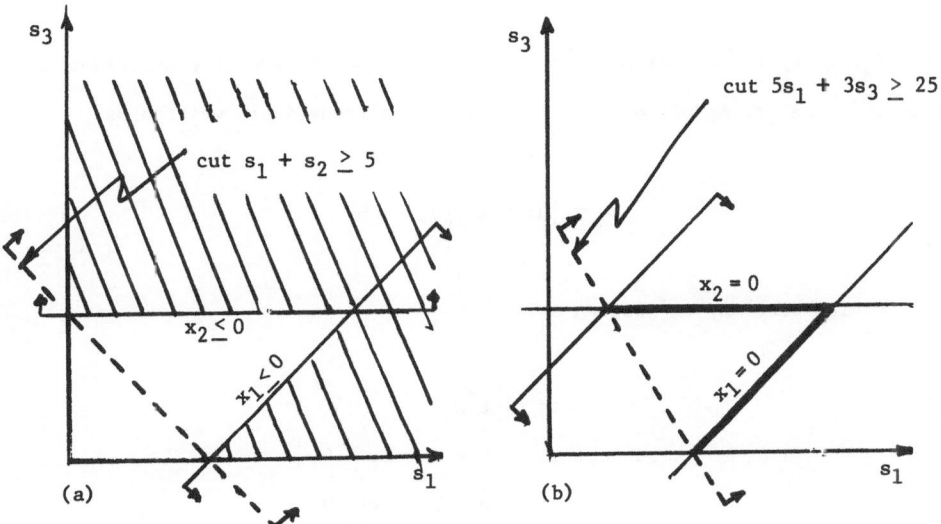

Figure 4.4. Illustration in the Nonbasic Variable Space

Through this example, we have demonstrated that one needs to consider all the constraints of the problem if a deepest cut is to be recovered through the disjunctive principles. However, this would lead to greater effort in the cut generation process. We now indicate the implication of this on the Generalized Lattice Point Problem, and the Linear Complementarity Problem.

4.3 Some General Comments with Applications to the Generalized Lattice Point and the Linear Complementarity Problems

Very briefly, we will illustrate through the Generalized Lattice Point and the Linear Complementarity problems the tradeoff which may be involved in the formulation of disjunctions. Recall our formulation of the Generalized Lattice Point Problem (GLPP) in Section 1.2.1 and let us rewrite it in a slightly different manner.

Note that if one selects q of the p components of u corresponding to linearly independent rows of A and restricts these q components to be zero, then one is confining the point u to some p-q dimensional facet of the set

$$U = \{u: u_i \geq 0, \ i=1,\ldots,p\} \qquad (4.9)$$

Letting $F_1,\ldots,F_{\hat{h}}$ be the set of such (p-q) dimensional facets of U and denoting $H=\{1,\ldots,\hat{h}\}$, the set S_h of Equation (1.6) may be alternatively written as

$$S_h = \{u: u \in F_h\}, \ h \in H \qquad (4.10)$$

Hence, Problem GLPP may be written as

$$
\begin{aligned}
\text{minimize} \quad & c^t x \\
\text{subject to} \quad & v = d - Dx \geq 0 \\
& u = b - Ax \geq 0 \\
& u \in \bigcup_{h \in H} S_h
\end{aligned}
$$

where S_h, $h \in H$ is given by Equation (4.10). If the rank of A is less than the number of rows A, that is, if A is not of full rank, then Problem GLPP poses an extra difficulty. One may relax the disjunctive statement in such a case to read

$$\{\text{at least q of the p components of u are zero}\} \qquad (4.11)$$

Now, one may unambiguously let $H=\{1,\ldots,\binom{p}{q}\}$ and correspondingly define sets S_h, $h \in H$, each corresponding to a unique combination of q out of p components of u restricted to zero. This modification makes the cut generation process much simpler and hence faster. However, this is at the expense of the depth of cut that can be derived therefrom.

Similarly, in the linear complementarity problem LCP considered in Section 1.2.5, one may be considering the violation of a particular disjunction $x_p x_q = 0$, say, in a solution to a relaxation of this problem. Hence, as exhibited in the foregoing section, one may simply use the constraints corresponding to $x_p \leq 0$ and $x_q \leq 0$ in order to derive a cut. Alternatively, one may choose to incorporate in the disjunction formulation the nonnegativity restrictions on the other basic variables as well.

Let us now generalize this concept to the situation of interest to us, namely the case where the objective function f is quasiconcave, and the set X of Problem DP (Section 1.1) is polyhedral. In order to establish connections between disjunctive programming techniques and polyhedral annexation methods, and to simplify the presentation, we will initially assume that each of the sets S_h, $h \in H$ is comprised of only a single constraint. Again, we will assume that a relaxation strategy is being adopted to solve Problem DP, so that currently, we have an extreme point optimal solution to the problem minimize $\{f(x): x \in X, x \geq 0\}$, which violates the disjunction $x \in \bigcup_{h \in H} S_h$. Here, we are assuming that the set X is comprised of the original linear constraints along with any valid inequalities which may have been generated over previous iterations. Accordingly, in terms of the current nonbasic variables, let the sets X and S_h, $h \in H$ be given by

$$X = \{x: Gx \leq g\} \equiv \{x: \sum_{j \in J} g_{ij} x_j \leq g_i \text{ for } i=1,\ldots,m\} \tag{4.12}$$

$$S_h = \{x: \sum_{j \in J} a_j^h x_j \geq 1, x \geq 0\}, h \in H \tag{4.13}$$

where J is the index set of the nonbasic variables. For each set S_h, $h \in H$, we have normalized the single constraint by its respective right-hand-side which must be positive since the origin violates each such constraint. Now, in order to derive a valid inequality which deletes the origin, one may invoke the disjunction

$$x \in \bigcup_{h \in H} S_h \tag{4.14}$$

However, we propose to derive stronger cuts by invoking the alternate disjunction

$$x \in \bigcup_{h \in H} XS_h \tag{4.15}$$

where,

$$XS_h \equiv X \cap S_h = \{x: \ Gx \leq g, \ \sum_{j \in J} a_j^h x_j \geq 1, \ x \geq 0\} \tag{4.16}$$

Note that one may invoke other valid disjunctions between the extremes (4.14) and (4.15) by adding on a subset of the constraints of X to each of the sets S_h, $h \in H$. As we have see, in the formulation of the disjunction, there is a tradeoff involved between the strength of the inequalities derived and the effort expended in generating these inequalities. Now, one viable approach is to commence with the disjunction (4.14) to obtain an initial cut, and then to sequentially add on constraints of X, attempting at each step to improve the current cut. This is basically the central point of the discussion of the following section.

4.4 Sequential Polyhedral Annexation

In this section, we will first briefly discuss the polyhedral annexation technique as is relevant to the present exposition. We will then demonstrate how an algorithmic scheme called sequential polyhedral annexation may be implemented to use the set X defined by (4.12) in order to improve the fundamental cut (3.19) available from the disjunction (4.14). We will also indicate some drawbacks of this method which lead us to proposing a variation of the scheme. This variation, however, involves additional effort in generating a cut.

Let us begin our discussion by making the observation that a disjunction which stipulates that at least one of the sets S_h of Equation (4.13) must be satisfied is equivalent to the statement that the interior of the polyhedron

$$S_H = \{x: \sum_{j \in J} a_j^h x_j \leq 1, \text{ for each } h \in H\} \qquad (4.17)$$

contains no feasible points in the nonnegative orthant. Henceforth, for the sake of convenience, we will call a polyhedron NFIP if its interior contains no feasible points in the nonnegative orthant. Thus, the polyhedral annexation procedure essentially does the following. Given several NFIP polyhedra, the technique suitably annexes them to each other in order to derive a new NFIP polyhedron of the type (4.17). Then, based on the constraints of this polyhedron, a cut of the type (3.19) is generated. The annexation scheme is based on the following main result.

Theorem 4.1

Let the polyhedra

$$S_P = \{x: \sum_j a_j^P x_j \leq b^P \quad \text{for each } p \in P\} \qquad (4.18)$$

and

$$S_Q = \{x: \sum_j a_j^q x_j \leq b^q \quad \text{for each } q \in Q\} \qquad (4.19)$$

be NFIP. Then, for any $k \in P$, and for any nonnegative parameters μ_{kq}, μ_q, $q \in Q$, the following polyhedron is NFIP:

$$S_R = \{x: \sum_j a_j^r x_j \leq b^r \text{ for each } r\epsilon R\}$$

$$\equiv \{x: \sum_j a_j^p x_j \leq b^p \text{ for each } p\epsilon P - \{k\} \tag{4.20}$$

$$\sum_j (\mu_{kq} a_j^k + \mu_q a_j^q) x_j \leq (\mu_{kq} b^k + \mu_q b^q) \text{ for each } q\epsilon Q\}$$

Proof. By contradiction, suppose S_R is not NFIP. Then, there exists a feasible, nonnegative x satisfying

$$\sum_j a_j^p x_j < b^p \quad \text{for each } p \epsilon P - \{k\}$$

$$\sum_j (\mu_{kq} a_j^k + \mu_q a_j^q) x_j < \mu_{kq} b^k + \mu_q b^q \text{ for each } q\epsilon Q$$

The first of these inequalities implies that $\sum_j a_j^k x_j \geq b^k$ or else, S_P would not be NFIP. This along with the second inequality implies that $\mu_q\{\sum_j a_j^q x_j - b^q\} < \mu_{kq}\{b^k - \sum_j a_j^k x_j\} \leq 0$, or that S_Q is not NFIP, a contradiction. This completes the proof.

In terms of the traditional disjunctive programming methods, Thoerem 4.1 has the following interpretation. The condition that at least one of the constraint sets

$$S_P = \{x: \sum_j a_j^p x_j \geq b^p, x \geq 0\}, p\epsilon P \tag{4.21}$$

and at least one of the constraint sets

$$S_q = \{x: \sum_j a_j^q x_j \geq b^q, x \geq 0\}, q\epsilon Q \tag{4.22}$$

must be satisfied, implies the weaker condition that at least one of the following constraint sets must be satisfied for some $k\epsilon P$

$$S_p \text{ for } p \epsilon P - \{k\},$$

$$\tag{4.23}$$

$$S_{k,q} = \{x: \sum_j a_j^k x_k \geq b^k, \sum_j a_j^q x_j \geq b^q, x \geq 0\} \text{ for } q \epsilon Q$$

Given any set of nonnegative surrogate multipliers μ_{kq}, μ_q for the two constraints in each of the sets $S_{k,q}$, $q \epsilon Q$, this in turn implies that at least one of the constraint sets

$$S_p \text{ for } p \epsilon P - \{k\},$$

$$\tag{4.24}$$

$$S_{kq} = \{x: \sum_j (\mu_{kq} a_j^k + \mu_q a_j^q) x_j \geq (\mu_{kq} b^k + \mu_q b^q), x \geq 0\} \text{ for } q \epsilon Q$$

must be satisfied, or that S_R of Equation (4.20) must be NFIP.

Clearly, the choice of $k \epsilon P$ for the purpose of annexation is crucial with regard to the strength of the inequality which may be derived from the disjunction (4.24). We will now discuss this choice in the context of a method known as sequential polyhedral annexation, as applied to the concepts introduced in Section 4.2.

Thus, suppose one has derived the following cut (3.19) from the disjunction that at least one of the sets S_h, $h \in H$ of Equation (4.13) must be satisfied

$$\sum_{j \epsilon J} \bar{\pi}_j x_j \geq 1 \tag{4.25}$$

The question addressed at this point is whether or not a given cut coefficient $\bar{\pi}_k$, $k \epsilon J$ can be improved (decreased) without worsening (increasing) the other coefficients. (In the discussion below, the reader may note that the sets X, S_H, S_P and S_Q are defined by (4.12), (4.17), (4.18) and (4.19) respectively). The manner in which the sequential method proposes to accomplish this is to commence with the NFIP polyhedron S_H and annex constraints of X one at a time. During this annexation process, that constraint which is a "blocking hyperplane", i.e., forms a "block", for the kth edge extension, is chosen to be surrogated with the newly added constraints. That is the cut coefficient $\bar{\pi}_k$ is determined by that particular constraint through (3.19). In other words, the surrogation serves the purpose of attempting to rotate this blocking hyperplane so as to permit an improved edge intercept. Of course, if more than one constraint form a block for the kth edge extension, then this process will have to be repeated for each of the blocking hyperplanes. Thus, starting with S_p equal to S_H, a set S_Q with $|Q| = 1$ is chosen to contain a single constraint of X. Let us assume that a constraint kεP of S_p forms a block for the kth edge extension. Then, S_p and S_Q are annexed through nonnegative parameters μ_{kq} and μ_q as follows.

Note that since the origin is infeasible to each S_p, pεP of Equation (4.21), we may assume as before without loss of generality that $b^p = 1$, pεP. To maintain consistency, we may also stipulate without loss of generality that the surrogation makes the right hand side of the constraint in S_{kq} of Equation (4.24) equal to unity, i.e., $\mu_{kq} + \mu_{kq}b^q = 1$. Thus, under the restriction that the cut derived from the disjunction (4.24) improves the kth edge intercept without worsening the other edge intercepts, we are searching for parameters μ_{kq}, μ_q satisfying

$$\mu_q \geq 0, \ \mu_{kq} = 1 - \mu_q b^q \geq 0 \tag{4.26}$$

$$\bar{\pi}_j \geq \mu_{kq}a_j^k + \mu_q a_j^q \quad \text{for each } j\varepsilon J \tag{4.27}$$

One may easily deduce from this that the appropriate choice reduces to finding the largest $\mu_q \geq 0$ satisfying

$$\mu_q \leq \underset{j\varepsilon J}{\text{minimum}} \left\{ \frac{\bar{\pi}_j - a_j^k}{(a_j^q - a_j^k b^q)} : (a_j^q - a_j^k b^q) > 0 \right\}$$

and $\tag{4.28}$

$$\mu_q b^q \leq 1$$

Then μ_{kq} is given through (4.26) and thus, the resulting NFIP polyhedron S_R of Equation (4.20) becomes the new polyhedron of the type (4.18). The cut (3.19) is updated, if necessary, with this new NFIP polyhedron (or disjunction) and the process is similarly repeated until the improvement of all edge intercepts have been attempted using all the constraints of X one at a time. Note that at each annexation, if the corresponding parameter μ_q obtained through (4.28) turns out to be zero, then this implies that $S_R \equiv S_P$ so that no improvement is possible with the current annexation.

Now, there is one principal drawback of this technique and that is, the final cut derived is dependent on the order in which one considers the constraints of X of Equation (4.12) to be used as sets S_Q of Equation (4.19). We illustrate this fact below through an example and then proceed to propose an alternative method.

Illustrative Example

Let us modify the example of Section 4.2 by adding an additional constraint to the set X of Equation (4.12). Hence, let the sets of Equation (4.13) or (4.21) be

$$S_1 = \{(s_1, s_3): \frac{s_1}{5} - \frac{s_3}{5} \geq 1, \; s_1, s_3 \geq 0\},$$

$$S_2 = \{(s_1, s_3): \frac{s_3}{5} \geq 1, \; s_1, s_3 \geq 0\} \qquad (4.29)$$

and suppose X is given by

$$X = \{(s_1, s_3): -s_1 + s_3 \leq 3, \; -s_1 + 3s_3 \leq 12\} \qquad (4.30)$$

The sets XS_1 and XS_2 of Equation (4.16) as well as the best cut available from the disjunction (4.15) are depicted in Figure 4.5.

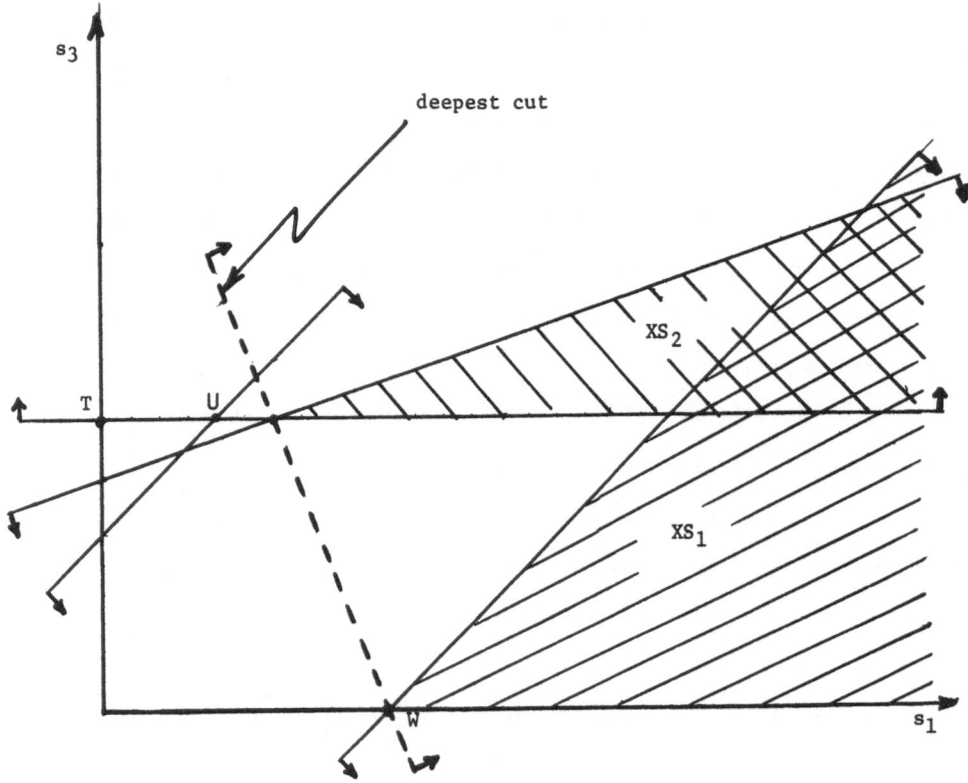

Figure 4.5. Deepest Cut From the Improved Formulation

Now, the cut (3.19) available from the disjunction $(s_1, s_3) \in S_1 \cup S_2$ is $\frac{s_1}{5} + \frac{s_3}{5} \geq 1$. This cut passes through the points T and W of Figure 4.5. One can see that the extension corresponding to edge s_1 cannot be improved. Hence, let us attempt to improve the edge intercept corresponding to s_3 using the sequential polyhedral annexation scheme. Towards this end, note that the constraint of S_2 represents the blocking hyperplane. Using the first constraint of X in the initial set S_Q of Equation (4.19), (with the inequality reversed) the relationships (4.28) yield

$$\mu_q \le \text{minimum}\left(\frac{\frac{1}{5} - 0}{1 - (0)(-3)} \ , \ \cdot\right), \ -3\mu_q \le 1, \ \mu_q \ge 0$$

The largest μ_q satisfying this is $\mu_q = \frac{1}{5}$, whence (4.26) gives $\mu_{kq} = 1 - (\frac{1}{5})(-3) = \frac{8}{5}$. Thus, the disjunction (4.24) is $(s_1,s_3) \in S_1 \cup S_{2q}$ where,

$$S_{2q} = \{(s_1,s_3): \frac{1}{5} s_1 + \frac{3}{25} s_3 \ge 1, \ s_1,s_3 \ge 0\} \equiv \text{New } S_2, \text{ say} \qquad (4.31)$$

The cut (3.19) from this disjunction is

$$\frac{1}{5} s_1 + \frac{3}{25} s_3 \ge 1 \qquad (4.32)$$

which passes through points U and W in Figure 4.5, and is also shown in Figure 4.4(b). Now let us repeat this by taking S_1 as in (4.29), S_2 as given by (4.31), the second constraint of X forming the set S_Q, and the constraint of S_2 representing the blocking hyperplane for the edge s_3 in the cut (4.32). The relationships (4.28) yield

$$\mu_q \le \text{minimum}\left(\frac{\frac{1}{5} - \frac{1}{5}}{1 - (\frac{1}{5})(-12)} \ , \ \cdot\right), \ -12\mu_q \le 1, \ \mu_q \ge 0$$

which implies, that $\mu_q = 0$ or that no further improvement is possible.

In this example, if one had considered the constraints of X in the reverse order then one would have obtained the deepest cut as shown in Figure 4.5. However, the appropriate ordering of the constraints of X is a combinatorial problem. Furthermore, conceivably it may be possible in some instances that the best cut is not recoverable no matter in which order the constraints of X are considered.

The method we propose to employ in the next section considers all the constraints of X simultaneously, that is, examines the disjunction (4.15) itself in an attempt to improve edge intercepts one at a time, holding other edge inter-

cepts fixed at each stage. This technique is easy to implement and directly
yields the best cut coefficients, the corresponding appropriate surrogate multipliers being available, if required, as a set of optimal dual variables.

4.5 A Supporting Hyperplane Scheme for Improving Edge Extensions

Suppose as before that we are given sets S_h, $h \in H$ defined by Equation
(4.13) with the stipulation that at least one of these sets must be satisfied.
We re-emphasize here that we continue to assume that each set S_h has only one
constraint merely for convenience. In addition, we are given a constraint set X
(Equation (4.12)) which must also be satisfied by an feasible point. The disjunction under consideration is that $x \in \bigcup_{h \in H} XS_h$ (Equation (4.15)) where, as in
Equation (4.16), $XS_h = X \cap S_h$, $h \in H$.

Thus, assume that currently, we have a cut of the form

$$\sum_{j \in J} \bar{\pi}_j x_j \geq 1 \tag{4.33}$$

which is valid for the disjunction (4.15). Note that initially, (4.33) may be
taken as the cut (3.19) derived from the disjunction $x \in \bigcup_{h \in H} S_h$.

Now, consider a $k \in J$ and suppose that we are presently trying to improve the
k^{th} edge intercept, that is, decrease $\bar{\pi}_k$. Towards this end, let us assume that
we are able to solve for each $h \in H$

$$P_{kh}: \quad \text{minimize} \quad \pi_{kh}$$

$$\text{subject to} \quad \pi_{kh} x_k + \sum_{\substack{j \in J \\ j \neq k}} \bar{\pi}_j x_j \geq 1 \text{ for each } x \in XS_h$$

$$\text{and} \tag{4.34}$$

$$\pi_{kh} x_k + \sum_{\substack{j \in J \\ j \neq k}} \bar{\pi}_j x_j = 1 \text{ supports } XS_h$$

Let

$$\bar{\pi}_k^* = \underset{h \in H}{\text{maximum}} \ \{\bar{\pi}_{kh}\} \qquad (4.35)$$

where $\bar{\pi}_{kh}$ is the solution to problem P_{kh}. Now consider the cut

$$\bar{\pi}_k^* x_k + \sum_{\substack{j \in J \\ j \neq k}} \bar{\pi}_j x_j \geq 1 \qquad (4.36)$$

Clearly, (4.36) is satisfied by each $x \in \underset{h \in H}{\cup} XS_h$, that is, (4.36) is a valid cut
for the disjunction (4.15). Moreover, any inequality $\sum_{j \in J} \pi_j x_j \geq 1$ with $\pi_j = \bar{\pi}_j$
for $j \in J - \{k\}$ and $\pi_k < \bar{\pi}_k^*$ is not valid because it deletes a point $\hat{x}$ of $XS_{\hat{h}}$ at
which the corresponding hyperplane $\bar{\pi}_{k\hat{h}} x_k + \sum_{\substack{j \in J \\ j \neq k}} \bar{\pi}_j x_j = 1$ supports $XS_{\hat{h}}$, where $\hat{h} \in H$
is an index for which equality holds in (4.35). To see this, it is sufficient
to show that if $\bar{\pi}_{kh} > -\infty$ in (4.34), then a point of support referred to in (4.34)
occurs at an $\bar{x}$ satisfying $\bar{x}_k > 0$. This is clearly so, for if not, then $\bar{\pi}_{kh}$ can
be reduced still further. Thus, (4.36) gives the best intercept possible for
the kth edge when all other intercepts are held fixed. Hence, replacing $\bar{\pi}_k$ of
(4.33) by $\bar{\pi}_k^*$, we would obtain a (possibly) new valid cut (4.33). This process
may now be repeated for each edge in turn till no further improvement is possible.
Of course, different cuts may be obtained by considering the edges in different
orders, but each of these cuts cannot be uniformly dominated by any other cut.

We will now proceed to discuss the determination of $\bar{\pi}_{kh}$, the coefficient
of x_k in the cutting plane under consideration, given through (4.34). The
problem we formulate below to accomplish this, has the following motivation.
Observe that the cut hyperplane is constrained to pass through (n-1) linearly
independent points of the form $(0, \ldots, \frac{1}{\pi_j}, \ldots, 0)$ for $j \in J - \{k\}$. In order to
uniquely define the cutting plane, we need to identify a suitable point $\bar{x}$ which
has $\bar{x}_k > 0$. Now, according to Equation (4.34), this cutting plane will need to
support the set XS_h with each point of XS_h being feasible to it. Hence, in order
to determine $\bar{\pi}_{kh}$, we may hold the intercepts on the axes $j \in J - \{k\}$ fixed and

decrease the intercept on the kth axis (increase π_{kh}) until the hyperplane merely supports XS_h at some point $\bar{x}$ with $\bar{x}_k > 0$. This problem is mathematically stated below. Theorem 4.2 later establishes that an optimal solution to this problem yields $\pi_k = \bar{\bar{\pi}}_{kh}$

$$\bar{P}_{kh}: \qquad \text{maximize} \qquad \pi_k$$

$$\text{subject to} \qquad \pi_k x_k + \sum_{\substack{j \in J \\ j \neq k}} \bar{\pi}_j x_j = 1 \tag{4.37}$$

$$x \in XS_h \tag{4.38}$$

$$x_k > 0 \tag{4.39}$$

Note that π_k is unrestricted in sign. Now using Equations (4.12), (4.13), (4.16) and solving for π_k through Equation (4.37), we may rewrite the above problem as

$$\text{maximize} \qquad \frac{1}{x_k} - \sum_{\substack{j \in J \\ j \neq k}} \bar{\pi}_j \left(\frac{x_j}{x_k} \right)$$

$$\text{subject to} \qquad \sum_{j \in J} g_{ij} \left(\frac{x_j}{x_k} \right) \leq \frac{g_i}{x_k} \qquad \text{for } i=1,\ldots,m$$

$$\sum_{j \in J} a_j^h \left(\frac{x_j}{x_k} \right) \geq \frac{1}{x_k}$$

$$\left(\frac{x_j}{x_k} \right) \geq 0, \quad x_k > 0$$

Finally, letting

$$\xi = \frac{1}{x_k} \quad \text{and} \quad y_j = \frac{x_j}{x_k} \qquad \text{for each } j \in J \tag{4.40}$$

we obtain the following linear programming problem in $|J|$ variables

$\text{LP}_{kh}:$ maximize $z(\xi,y) = \xi - \sum_{\substack{j \in J \\ j \neq k}} \bar{\bar{\pi}}_j y_j$

subject to $\sum_{\substack{j \in J \\ j \neq k}} g_{ij} y_j - g_i \xi \leq -g_{ik}$ for $i=1,\ldots,m$

$\xi - \sum_{\substack{j \in J \\ j \neq k}} a_j^h y_j \leq a_k^h$

$\xi \geq 0, \; y_j \geq 0$ for $j \in J - \{k\}$

Consider the following result.

Theorem 4.2

If Problem LP_{kh} is feasible, then it has an optimal solution $\bar{\xi}, \bar{y}_j, \; j \in J - \{k\}$ with $\bar{\xi} < \infty$. Moreover, the optimal solution values of Problems LP_{kh} and Problem P_{kh} (defined by 4.34)) are equal.

Proof. Note that the constraints of Problem LP_{kh} may be rewritten as $\sum_{j \in J} g_{ij} y_j - g_i \xi < 0$ for $i=1,\ldots,m;$ $- \sum_{j \in J} a_j^h y_j + \xi \leq 0$ and $y_k = 1$, with $\xi, y \geq 0$. Letting $\mu_i, \; i=1,\ldots,m, \; \gamma$ and β_k be the respective dual variables associated with these constraints, the dual to Problem LP_{kh} may be written as

$\text{DLP}_{kh}:$ minimize β_k

subject to $a_j^h \gamma - \sum_{i=1}^{m} g_{ij} \mu_i \leq \bar{\pi}_j$ for $j \in J - \{k\}$ (4.41)

$a_k^h \gamma - \sum_{i=1}^{m} g_{ik} \mu_i \leq \beta_k$ (4.42)

$\gamma - \sum_{i=1}^{m} g_i \mu_i \geq 1$ (4.43)

$\gamma, \; \mu \geq 0$

Letting $\bar{\beta}_k$ denote the minimum value of β_k, we will show that $\bar{\beta}_k = \bar{\pi}_{kh}$. We have from (4.34)

$$\bar{\pi}_{kh}x_k + \sum_{\substack{j\epsilon J \\ j \neq k}} \bar{\pi}_j x_j \geq 1 \qquad \text{for each } x\epsilon XS_h \qquad (4.44)$$

$$\bar{\pi}_{kh}\hat{x}_k + \sum_{\substack{j\epsilon J \\ j \neq k}} \bar{\pi}_j \hat{x}_j = 1 \qquad \text{for some } \hat{x}\epsilon XS_h \qquad (4.45)$$

Hence (4.44) is implied by XS_h and Lemma 2.2 asserts the existence of $\hat{\gamma} \geq 0$, $\hat{\mu}_i \geq 0$ satisfying (4.41) through (4.43) with $\beta_k = \bar{\pi}_{kh}$. That is $\hat{\gamma}$, $\hat{\mu}_i$ and $\bar{\pi}_{kh}$ is feasible to DLP_{kh}. Thus LP_{kh} is bounded. Hence, $\bar{\beta}_k \leq \bar{\pi}_{kh} < \infty$. Now let $\bar{\gamma}$, $\bar{\mu}_i$ and $\bar{\beta}_k$ solve DLP_{kh}. Then, $x\epsilon XS_h$ implies

$$\sum_{j\epsilon J} [a_j^h\bar{\gamma} - \sum_{i=1}^m g_{ij}\bar{\mu}_i]x_j \geq [\bar{\gamma} - \sum_{i=1}^m g_i\bar{\mu}_i]$$

Then noting (4.41) through (4.43), we get

$$\bar{\beta}_k\hat{x}_k + \sum_{\substack{j\epsilon J \\ j \neq k}} \bar{\pi}_j \hat{x}_j \geq 1 \qquad \text{for each } x\epsilon XS_h \qquad (4.46)$$

We have shown that $\bar{\beta}_k \leq \bar{\pi}_{kh}$. Now if $\bar{\beta}_k < \bar{\pi}_{kh}$, then from (4.45)

$$\bar{\beta}_k x_k + \sum_{\substack{j\epsilon J \\ j \neq k}} \bar{\pi}_j x_j < 1$$

contradicting (4.46). Hence $\bar{\beta}_k = \bar{\pi}_{kh}$.

Finally, since LP_{kh} is bounded, there exists an optimal extreme point solution $(\bar{\xi},\bar{y})$ with $\bar{\xi}$ finite. This completes the proof.

Corollary

Let $\bar{\xi}$, $\bar{y}_j$, $j\epsilon J - \{k\}$ solve LP_{kh} with $\bar{\xi} < \infty$, and with $\bar{\beta}_k$ as the corresponding objective function value. Then, $\pi_k = \bar{\beta}_k$, $x_k = \frac{1}{\bar{\xi}}$ and $x_j = \frac{\bar{y}_j}{\bar{\xi}}$ for $j\epsilon J - \{k\}$ solves

Chapter V

GENERATION OF FACETS OF THE CLOSURE OF THE
CONVEX HULL OF FEASIBLE POINTS

5.1 Introduction

In this Chapter, we examine a procedure for replacing the disjunctive

statement in a problem by linear inequalities which represent the facets of the

closure of the convex hull of points feasible to the disjunction. In particular,

we present necessary and sufficient conditions for an inequality to define a

facet of the closure of the convex hull of feasible points. The actual generation

of such facets is a hard problem. However, for a special class of problems

(called "facial problems" in this chapter) it is possible to obtain the closure

of the convex hull of points satisfying disjunctions, in a sequence of q steps,

where each step generates the closure of the convex hull of points satisfying

one disjunction only.

To simplify the presentation, we will avoid proving results and will simply

state them and illustrate them through a numerical example. To begin with, let

us state the form of the disjunctive program DP of Chapter I which we will be

working with in the present context

$$\underline{DP}: \qquad \text{minimize} \quad f(x) = c^t x$$
$$\text{subject to} \quad x \in X = \{x: Dx \geq d, \ x \geq 0\}$$
$$\underset{h \in H}{V} \ \{A^h x \geq b^h, \ x \geq 0\}$$

The linear program obtained by relaxing the disjunctive statement of Problem DP is

$$\underline{LP}: \qquad \text{minimize} \quad c^t x$$
$$\text{subject to} \quad Dx \geq d$$
$$x \geq 0$$

We will assume henceforth that both DP and LP are stated above in terms of the nonbasic variables at the current optimal solution to LP. Thus, the current solution is $x = 0$ with feasibility implying that $d \leq 0$ and optimality implying that $c \geq 0$.

To illustrate, let us work with the numerical problem of Section 4.2 throughout this chapter. From the optimal tableau for Problem LP given in Section 4.2, we deduce

$$
\begin{array}{lll}
\text{DP:} & \text{minimize} & 2s_1 + s_3 \\
& \text{subject to} & -s_1 + s_3 \geq -5 \\
& & s_1 - s_3 \geq -3 \\
& & - s_3 \geq -5 \\
& & s_1, s_3 \geq 0
\end{array}
$$

$$
\left\{ \begin{array}{c} s_1 - s_3 \geq 5 \\ s_1, \ s_3 \geq 0 \end{array} \right\} \ \bigvee \ \left\{ \begin{array}{c} s_3 \geq 5 \\ s_1, \ s_3 \geq 5 \end{array} \right\}
$$

5.2 A Linear Programming Equivalent of the Disjunctive Program

In this section, we will write a linear program ELP which is in a defined sense equivalent to the nonconvex problem DP given above. For this purpose let us define the following sets.

For each $h \in H$, let

$$
F_h = \{x \in R^n: D^h x \geq d^h, \ x \geq 0\} \equiv \{x \in R^n: Dx \geq d, \ A^h x \geq b^h, \ x \geq 0\} \tag{5.1}
$$

represent the points feasible to LP which are also feasible to the hth, $h \in H$, disjunctive constraint. Also let

$$
F = \bigcup_{h \in H} F_h = \left\{ x \in R^n: \bigvee_{h \in H} \{D^h x \geq d^h, \ x \geq 0\} \right\} \tag{5.2}
$$

Finally, let us denote the feasible region of LP as

$$F_0 = \{x \epsilon R^n: Dx \geq d, \ x \geq 0\} \tag{5.3}$$

Let us assume that $|H| < \infty$ and let us define

$$H^* = \{h \epsilon H: F_h \neq \{\phi\}\} \tag{5.4}$$

Now, let us characterize the closure of the convex hull of F, denoted by
<u>cl conv F</u>. Note that any $x \epsilon F$ may be written as

$$x = \sum_{h \epsilon H^*} \xi_0^h u^h, \text{ where } u^h \epsilon F_h, \ h \epsilon H^*, \text{ and where,}$$

$$\sum_{h \epsilon H^*} \xi_0^h = 1, \ \xi_0^h \geq 0$$

Hence, substituting $\xi^h = \xi_0^h u^h$ and noting that $D^h u^h \geq d^h$, $u^h \geq 0$, we get

$$cl \ conv \ F = \left\{ \begin{array}{l} x \epsilon R^n: x = \sum_{h \epsilon H^*} \xi^h \\[2ex] D^h \xi^h \geq d^h \xi_0^h, \ h \epsilon H^* \\[2ex] \sum_{h \epsilon H^*} \xi_0^h = 1 \\[2ex] (\xi_1^h, \ \xi_0^h) \geq 0, \ h \epsilon H^* \end{array} \right\} \tag{5.5}$$

It may be shown that if the feasible region of LP is bounded, then (5.5) is true
with H* replaced by H. Now, if LP has a finite optimal solution, then directly
using the characterization of Equation (5.5), we may write a linear program ELP
equivalent to Problem DP in the sense of Theorem 5.1 stated below.

ELP: minimize $\sum\limits_{h\varepsilon H} c^t \xi^h$

subject to $D^h \xi^h \geq d^h \xi_0^h$, $h\varepsilon H$

$\sum\limits_{h\varepsilon H} \xi_0^h = 1$

$(\xi^h, \xi^h) \geq 0$, $h\varepsilon H$

Let us denote the feasible region of Problem ELP as P. Then consider the following result stated without proof.

Theorem 5.1

Problems DP and ELP are equivalent in the following sense

(i) For every extreme point x of $c\ell$ conv F, there corresponds an extreme point of P with components

$$\begin{cases} (\xi^k, \xi_0^k) = (x, 1) \text{ for some } k\varepsilon H \\ \\ 0 \quad \text{otherwise} \end{cases}$$

(ii) All extreme points of P have components of the following form

$$\begin{cases} (\xi^k, 1) \qquad \text{for some } k\varepsilon H \\ \\ 0 \quad \text{otherwise} \end{cases}$$

where, $x = \xi^k$ is an extreme point of F_k

(iii) x is optimal to DP if and only if the corresponding extreme point of P defined in (i) above is optimal to ELP.

For the moment, we will not involve ourselves with the description of specialized solution procedures for Problem ELP. We merely remark that there do exist simplex-based decomposition solution schemes which exploit the structure of ELP. We will now proceed to give an alternate characterization of the set $c\ell$ conv F which permits the explicit generation of the facets of the convex hull

of feasible points. Before that, let us illustrate some of the concepts introduced in this section through our numerical example. Figure 5.1 is provided
below for this purpose, and may be referred to along with the statement of
Problem DP.

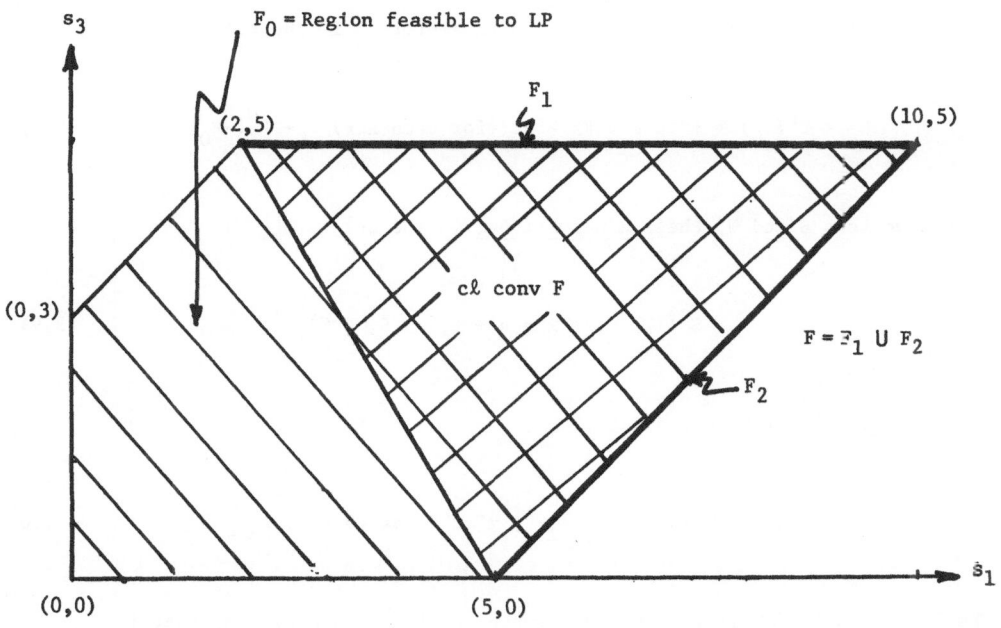

Figure 5.1. Illustration of the sets F_0, F_1, F_2, F, cℓ conv F

5.3 Alternative Characterization of the Closure of the Convex Hull of Feasible Points

In this section, we will lay the foundations for the procedure which will
generate facets of the set cℓ conv F. In particular, we will present an
alternative characterization for the latter set. To begin with, let us informally
or conceptually introduce certain definitions which we will find necessary to
use in the following exposition.

By the <u>polar set</u> of F, we mean the set

$$F^0 = \{\pi \in R^n : \pi x \leq 1 \text{ for each } x \in F\} \qquad (5.6)$$

By the _scaled polar_ of F with scale parameter π_0, we mean the set

$$F^0(\pi_0) = \{\pi \in R^n: \pi x \leq \pi_0 \quad \text{for each } x \in F\} \tag{5.7}$$

The _scaled reverse polar of F_ with scale parameter π_0 is the set

$$F^\#(\pi_0) = \{\pi \in R^n: \pi x \geq \pi_0 \quad \text{for each } x \in F\} \equiv - F^0(-\pi_0) \tag{5.8}$$

Given two sets S and T, their _Minkowski Sum_ is the set

$$S + T = \{x: x = s + t, \ s \in S, \ t \in T\} \tag{5.9}$$

The _conical hull_ of S is the set

$$\text{cone } S = \{x: x = \sum_i \lambda_i x^i, \ x^i \in S \text{ for each } i, \ \lambda_i \geq 0\} \tag{5.10}$$

The _linear hull_ of S is the smallest subspace of R^n containing S, that is,

$$\ell h \ S = \{x \in R^n: x = \sum_i \lambda_i x^i, \ x^i \in S \text{ for each } i\} \tag{5.11}$$

The _affine hull_ of S is the set

$$\text{aff } S = \{x \in R^n: x = \sum_i \lambda_i x^i, \ x^i \in S \text{ for each } i, \ \sum_i \lambda_i = 1\} \tag{5.12}$$

The _orthogonal complement_ of S is the set

$$S^\perp = \{\pi \in R^n: \pi x = 0 \text{ for each } x \in S\} \tag{5.13}$$

The _recession cone_ of S is the closed cone which is a closure of the set of directions of S, that is,

$$C(S) = c\ell\{v \in R^n: x\in S \text{ implies that } x + \lambda v \in S \text{ for all } \lambda \geq 0\} \qquad (5.14)$$

The <u>linearity</u> of S is the dimensionality of the largest subspace contained in C(S), that is,

$$\ell\text{in } S = \dim \{R: R \text{ is a subspace of } C(S) \text{ and}$$

$$Q \subset R \text{ for each subspace } Q \text{ of } C(S)\} \qquad (5.15)$$

Note that for a set T, the <u>dimension</u> of T, denoted <u>dim T</u>, is the dimension of the linear hull of T.

Let us turn our attention to the scaled reverse polar of F, $F^{\#}(\pi_0)$, (Equation (5.8)) which we will find very important in the present context. Note that $F^{\#}(\pi_0)$ is the set of all normals to the hyperplanes which define valid cuts for the stated disjunction. Hence, two alternative ways of writing $F^{\#}(\pi_0)$ are given below

(a) $F^{\#}(\pi_0) = \{\pi \in R^n: \pi x^i \geq \pi_0 \text{ for each } x^i \in \text{vert } c\ell \text{ conv } F$

$\qquad\qquad\qquad \pi d^i \geq 0 \quad \text{for each } d^i \in \text{dir } c\ell \text{ conv } F\}$ $\qquad (5.16)$

(b) $F^{\#}(\pi_0) = \{\pi \in R^n: \pi \geq \theta^h D + \sigma^h A^h, h\in H^* \text{ for some } \theta^h, \sigma^h \geq 0,$

$\qquad\qquad\qquad h\in H^* \text{ such that } \theta^h d + \sigma^h b^h \geq \pi_0\}$ $\qquad (5.17)$

Above, for a polyhedral set S, <u>vert S</u> denotes the set of extreme points of the set S and <u>dir S</u> denotes the set of directions of S.

Note that the sign of π_0 is important in this context since we can always scale $\pi x \geq \pi_0$ so that π_0 is either +1 or -1 or 0. These latter three cases will henceforth be of primary interest to us. Further, whenever the sign of π_0 is inconsequential, we will simply write $F^{\#}$ instead of $F^{\#}(\pi_0)$. Finally, note that the characterization (5.16) of $F^{\#}(\pi_0)$ is conceptual whereas that of Equation (5.17) is accessible. To aid our understanding of $F^{\#}$, let us actually construct it using (5.16) for our problem, using values of 1, -1, and 0 for π_0.

As depicted in Figure 5.1, $c\ell$ conv F is a polytope with extreme points (2,5), (5,0), (10,5). The set dir $c\ell$ conv F is vacuous. Using Equation (5.16), we have,

$$F^{\#}(\pi_0) = \left\{ \begin{array}{l} \pi = (\pi_1,\pi_2): \ 2\pi_1 + 5\pi_2 \geq \pi_0 \\[2ex] 5\pi_1 \geq \pi_0 \\[2ex] 10\pi_1 + 5\pi_2 \geq \pi_0 \\[2ex] \pi_1, \ \pi_2 \ \text{unrestricted} \end{array} \right\}$$

Figure 5.2 illustrates the sets $F^{\#}(1)$, $F^{\#}(-1)$ and $F^{\#}(0)$.

Some useful properties of the reverse polar are stated below without proof.

Lemma 5.1

Let S and T be arbitrary sets. Then,

$$(\lambda S)^{\#} = \frac{1}{\lambda} (S^{\#}), \ -\infty < \lambda < \infty$$

$$S \subseteq T \text{ implies } S^{\#} \supseteq T^{\#}$$

$$(S \cup T)^{\#} = S^{\#} \cap T^{\#}$$

Based on these properties, one may establish the following important relationships between $F^{\#}$ and $c\ell$ conv F.

Theorem 5.2

(i) If $\pi_0 > 0$, then

$$0 \ \varepsilon \ c\ell \text{ conv } F \ <=> \ F^{\#} = \{\phi\} \ <=> \ F^{\#} \text{ is bounded}$$

(ii) If $\pi_0 \leq 0$, then $F^{\#} \neq \{\phi\}$ and

$$0 \ \varepsilon \ \text{int conv } F \ <=> \ F^{\#} \text{ is bounded.}$$

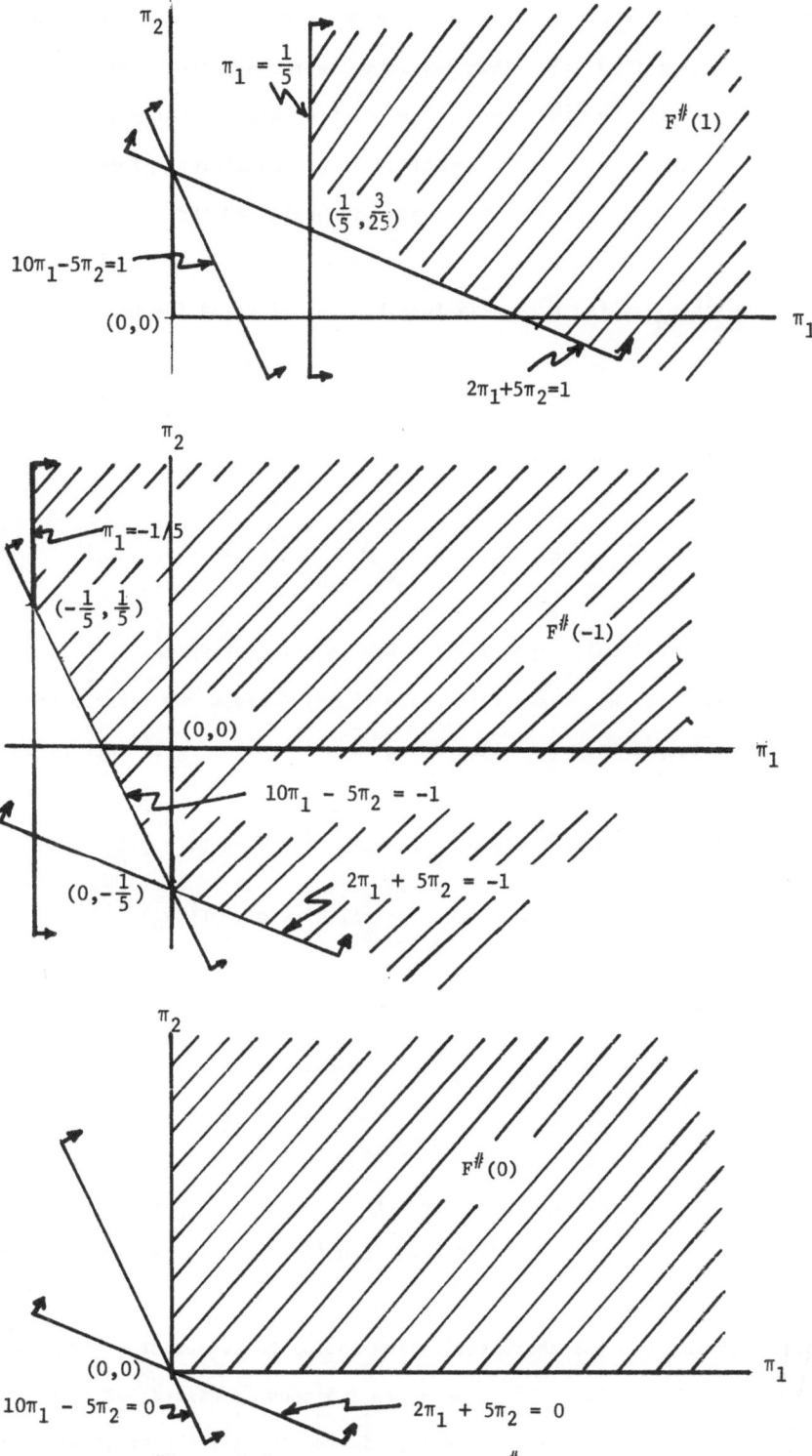

Figure 5.2. Construction of $F^{\#}(\pi_0)$ for $\pi_0 = 1, -1, 0$

However, the set more closely related with cl conv F is the set $(F^{\#})^{\#}$ denoted by $F^{\#\#}$. Suppose that $F^{\#} \neq \{\phi\}$. That is, from Theorem 5.2 above, assume that $0 \notin$ cl conv F. In other words, the point we are currently located at (assumed to be the origin) is infeasible to the disjunction. Then, note from Equation (5.8) and the definition of valid inequalities that

$$F^{\#\#}(\pi_0) = \{x \in R^n = x \text{ is feasible to all valid inequalities}$$
$$\text{of the form } \pi x \geq \pi_0\} \qquad (5.18)$$

Thus, clearly, we have,

$$\text{cl conv F} = \bigcap_{\pi_0 = 1, -1, 0} F^{\#\#}(\pi_0) \qquad (5.19)$$

Since the set of all valid inequalities are jointly determined by $F^{\#}(1)$, $F^{\#}(-1)$, $F^{\#}(0)$, then from (5.18), the intersection (5.19) defines the closure of the convex hull of feasible points. Two alternative ways of rewriting $F^{\#\#}$ are given below

(a) $$F^{\#\#}(\pi_0) = \left\{ x \in R^n \colon \begin{array}{l} \pi^1 x \geq \pi_0 \text{ for all } \pi^1 \in \text{vert } F^{\#}(\pi_0) \\[2mm] d^1 x \geq 0 \text{ for all } d^1 \in \text{dir } F^{\#}(\pi_0) \end{array} \right\} \qquad (5.20)$$

(b) From definitions (5.9), (5.10) and (5.19) one may alternatively show, that

$$F^{\#\#}(\pi_0) = \begin{cases} \text{cl (conv F + cone F)} & \text{if } \pi_0 > 0 \\ \text{cl cone F} & \text{if } \pi_0 = 0 \\ \text{cl conv (F } \cup \{0\}) & \text{if } \pi_0 < 0 \end{cases} \qquad (5.21)$$

As before, we may use the characterization (5.16) to construct $F^{\#\#}(\pi_0)$ from the sets $F^{\#}(\pi_0)$ of Figure 5.2. These sets are depicted in Figure 5.3 below. The reader may find it interesting to verify diagramatically the definition of Equation (5.21).

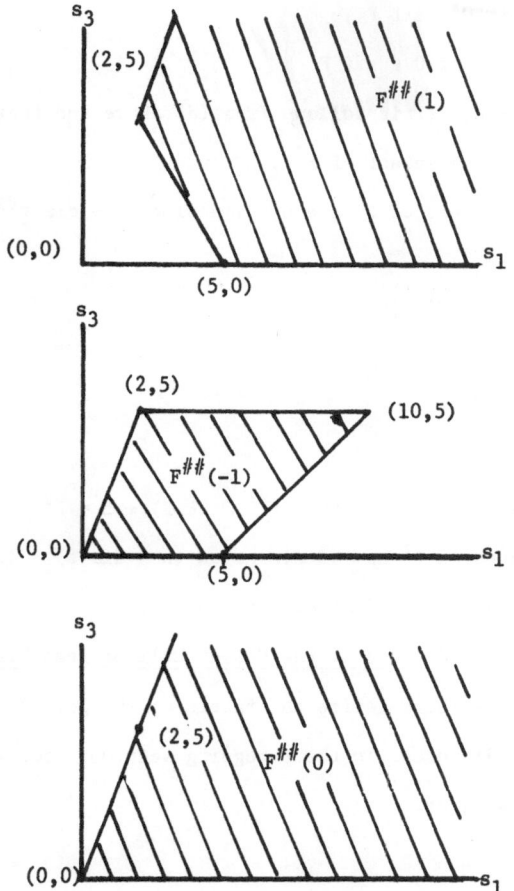

Figure 5.3. Construction of $F^{\#\#}(\pi_0)$ for $\pi_0 = 1,-1,0$

Next, we state some useful properties of the set $F^{\#}$ and some important results concerning $F^{\#}$ and $F^{\#\#}$. These are based on the definitions (5.8) through (5.15) introduced earlier.

Lemma 5.2

    (a) If $F^{\#} = \{\phi\}$ (which necessarily means that $\pi_0 > 0$), then $F^{\#\#} = R^n$

    (b) $F^{\#\#\#} = F^{\#}$

    (c) aff $F^{\#\#} = \ell h\ F^{\#\#} = \ell h\ F$

    (d) $\ell h\ F = L^{\perp}$ where

        L = largest subspace contained in the recession cone $C(F^{\#})$

          of $F^{\#} \equiv$ linearity space of $F^{\#}$ (5.22)

    (e) dim $F^{\#\#}$ + $\ell in\ F^{\#\#} = n$

(f) $\dim F^{\#\#} = \begin{cases} \dim F \text{ if } 0 \, \epsilon \text{ aff } F \\ \dim F + 1 \text{ if } 0 \, \cancel{\epsilon} \text{ aff } F \end{cases}$

(g) $F^{\#} = (F^{\#} \cap \ell h \; F) + L.$ (This follows from (d) above and fact that L

   defined in (5.22) is a subset of $F^{\#}$).

(h) Lowest dimensional faces of $F^{\#}$ are of dimension $(n - \dim F^{\#\#})$.

For our example problem,

   aff $F^{\#\#} \equiv \ell h \; F^{\#\#} \equiv \ell h \; F \equiv L^{\perp} \equiv R^2$

   $\ell in \; F^{\#} \equiv L = \{\phi\}$ for each π_0

   $\dim F^{\#\#} = 2 \; (=n)$

   Also, $0 \, \epsilon$ aff F and $\dim F = 2$

   Since $L = \{\phi\}$, $\ell h \; F = R^2$, $F^{\#} \subset R^2$, $F^{\#} = (F^{\#} \cap \ell h \; F) + L = F^{\#} \cap R^2$.

   Lowest dimensional faces of $F^{\#}$ are of dimension $(n - \dim F) = (2-2) = 0$

   i.e., lowest dimensional faces of $F^{\#}$ are extreme points.

5.4 Generation of Facets of the Closure of the Convex Hull of Feasible Points

In this section, we will characterize the facets of the set $c\ell$ conv F in
terms of the sets $F^{\#}$ and $F^{\#\#}$ discussed in the foregoing section. For the sake of
completeness, let us define a facet.

Definition

$\pi x \geq \pi_0$ is a __facet__ of a d-dimensional set S if

$\pi x \geq \pi_0$ for each $x \, \epsilon$ S and

$\pi x = \pi_0$ for exactly d affinely independent points of S.

Theorem 5.3 stated next characterizes the facets of the set $F^{\#\#}$ which play
an instrumental role in determining the facets of $c\ell$ conv F as will be seen shortly.

Theorem 5.3

$\pi x \geq \pi_0$ is a facet of $F^{\#\#}$ __and__ $\pi \; \epsilon \; \ell h \; F$

if and only if

$$\left\{ \begin{array}{l} \pi_0 \neq 0, \; \pi \neq 0, \; \pi \; \epsilon \; \text{vert} \; (F^{\#} \cap L^{\perp}) \\ \text{or} \;\; \pi_0 = 0, \; \pi \neq 0, \; \pi \; \epsilon \; \text{dir} \; (F^{\#} \cap L^{\perp}) \end{array} \right\}$$

Note: Recall from Lemma 5.2(d) that $L^{\perp} = \ell h \; F$. In this theorem, the statement
$\pi \; \epsilon \; \ell h \; F$ becomes necessary for the following technicality. Suppose $F^{\#\#}$ is less

than full dimensional. Then a facet of $F^{\#\#}$ defined by a hyperplane H can also be defined by a family of hyperplanes H', such that

$$(\ell h \ F) \cap H' = (\ell h \ F) \cap H$$

Thus, we specify a particular hyperplane from this family which has its normal π lying in ℓh F.

Since for our problem, the largest subspace contained in the recession cone of $F^{\#}$ is of zero dimension, i.e., $\ell in \ F^{\#} = 0$. Thus, (5.22) yields L = $\{\phi\}$ or that $\ell h \ F \equiv L^{\perp} = R^2$. Hence, $F^{\#} \cap L^{\perp} = F^{\#}$ in this case. Thus, in the present case, Theorem 5.3 above follows from the fact that $F^{\#\#}$ may be defined as in Equation (5.20).

Let us now consider an important result for computing facets of $F^{\#\#}$.

Theorem 5.4

Let $g \ \varepsilon \ R^n$, $\pi x \geq \pi_0$ be a facet of $F^{\#\#}$ and let $\pi \ \varepsilon \ \ell h$ F. Then, $g \ \varepsilon \ \{x \ \varepsilon \ F^{\#\#}: \ \pi x = \pi_0\} <=> \pi g = \pi_0$ supports $F^{\#}$ and contains π, where,

$$\pi \ \varepsilon \begin{cases} \text{vert } F^{\#} \cap L^{\perp} \text{ if } \pi_0 \neq 0 \\ \\ \text{dir } F^{\#} \cap L^{\perp} \text{ if } \pi_0 = 0 \end{cases}$$

For example, in our illustrative problem, consider $\pi_0 = 0$. Let us examine $F^{\#\#}(0)$. The facet $5s_1 - 2s_3 \geq 0$ has $\pi = (5,-2)$ lying in $\ell h \ F \equiv R^2$. Further, $g = (2,5)^t$ say, belongs to $F^{\#\#}(0)$ with $\pi g = 0$. (Note that $\pi_0 = 0$ here). Then the theorem implies that $2x_1 + 5x_2 = 0$ supports $F^{\#}(0)$ and contains the point $\pi = (5,-2)$ where, as seen in Figure 5.2, $\pi = (5,-2) \ \varepsilon \ \text{dir } F^{\#}(0)$ (since $L^{\perp} = R^2$ here). The converse result is also true.

Now, let us consider the main result of this section. This result characterizes the facets of $c\ell$ conv F in terms of the sets $F^{\#}$ and $F^{\#\#}$. Two cases are considered, e.g., $\pi_0 \neq 0$ and $\pi_0 = 0$. A discussion of this theorem and its implications follow after its statement.

Theorem 5.5

    (i) Suppose $\pi_0 \neq 0$. Further, assume that $0 \in$ aff F. Then,

$$\{\pi x \geq \pi_0 \text{ is a facet of } c\ell \text{ conv } F\} <=> \{\pi x \geq \pi_0 \text{ is a facet of } F^{\#\#}\}$$

Corollary

Assume $0 \in$ aff F, $\pi \in \ell h$ F. Then,

$$\{\pi x \geq \pi_0 \text{ is a facet of } c\ell \text{ conv } F\} <=> \{\pi \neq 0, \pi \in \text{vert } F^{\#} \cap L^{\perp}\}$$

    (ii) Suppose $\pi_0 = 0$. If $\pi x \geq 0$ is a facet of $c\ell$ conv F, $\pi \in \ell h$ F, then this implies that

$$\pi \neq 0, \text{ and that } \pi \text{ is an extreme direction of } F^{\#} \cap L^{\perp}$$

Conversely, if $\pi \neq 0$, $\pi \in$ dir $(F^{\#} \cap L^{\perp})$, and d=dim $F^{\#\#}$, then either

(a) $\pi x \geq 0$ is a facet of $c\ell$ conv F, or

(b) $\pi x \geq 0$ is a (d-2) dimensional face of $c\ell$ conv F such that this (d-2)

dimensional face is an intersection of two adjacent facets of the type

$$\pi^1 x \geq \pi_0^1, \ \pi_0^1 > 0$$

and

$$\pi^2 x \geq \pi_0^2, \ \pi_0^2 < 0$$

with

$$\pi = \left(\frac{\pi^1}{\pi_0^1} - \frac{\pi^2}{\pi_0^2} \right)$$

Note: Let us consider case (i) first, e.g., $\pi_0 \neq 0$. Note that if $0 \notin$ aff F, then this implies that dim F = d-1 and that each facet $\pi x = \pi_0$ of $F^{\#\#}$ contains all of cl conv F, instead of being a facet of cl conv F as when 0ε aff F. Thus, in case $0 \notin$ aff F, a facet of cl conv F is given by the intersection of $\pi x = \pi_0$ and a (d-2) dimensional face of $F^{\#\#}$ corresponding to an edge or a 1 dimensional face of $F^{\#} \cap L^{\perp}$.

Secondly, note from Lemma 5.2(c), (d) that if $F^{\#\#}$ is full dimensional, then $L = \{\phi\}$, or, $L^{\perp} = R^n$. This implies that one may examine the non-zero vertices of $F^{\#}$ itself to obtain facets of cl conv F.

Thirdly, if $d = $ dim $F^{\#\#} < n$, then $lin F^{\#} = (n-d) > 0$ (see Lemma 5.2 (e)). Hence, $F^{\#}$ has no extreme points. However, there is a one-to-one correspondence between vertices π of $F^{\#} \cap L^{\perp}$ and (n-d) dimensional faces of $F^{\#}$ which are of the form $\alpha + L$, where L is given by (5.22). These are the lowest dimensional faces of $F^{\#}$ (Lemma 5.2(h)).

Again, the corollary in case (i) of the theorem designates $\alpha + L$ to be an (n-d) dimensional face of $F^{\#}$. In particular, if $F^{\#\#}$ is of full dimension, i.e., d=n, then this Corollary states that

$$\{\pi x \geq \pi_0 \text{ is a facet of } cl \text{ conv F}\} \iff \{\pi \neq 0 \text{ is an extreme point of } F^{\#}\} \quad (5.23)$$

Case (ii) of the theorem states that if $\pi \neq 0$ belongs to the set dir $F^{\#} \cap L^{\perp}$ or, if $F^{\#\#}$ is of full dimension, and $\pi \neq 0$ belongs to dir $(F^{\#})$ itself, then $\pi x \geq 0$ may be used in a system of linear inequalities which characterize cl conv F in either case (a) or (b) of the converse.

Let us illustrate some aspects of this theorem through our numerical example first, and then discuss the utility of this theorem.

Let us begin with Case (i). Note that 0ε aff F and that $F^{\#} \cap L^{\perp} = F^{\#}$ in our problem. The facets of cl conv F of the type $\pi x \geq \pi_0$, $\pi_0 \neq 0$, are

(1) $5s_1 + 3s_3 \geq 25$, i.e., $\dfrac{s_1}{5} + \dfrac{3s_3}{25} \geq 1$, which is a facet of $F^{\#\#}(1)$ with

$\left(\dfrac{1}{5}, \dfrac{3}{25}\right)$ being the only extreme point of $F^{\#}(1)$

(2) $-s_1 + s_3 \geq -5$, i.e., $\dfrac{-s_1}{5} + \dfrac{s_3}{5} \geq -1$, which is a facet of $F^{\#\#}(-1)$ with

$\left(\dfrac{-1}{5}, \dfrac{1}{5}\right)$ being an extreme point of $F^{\#}(-1)$

and

(3) $-s_3 \geq -5$, i.e., $\dfrac{-s_3}{5} \geq -1$, which is also a facet of $F^{\#\#}(-1)$ with

$\left(0, \dfrac{-1}{5}\right)$ being the only other extreme point of $F^{\#}(-1)$.

To verify Case (ii), observe that extreme directions π of $F^{\#}(0) \cap L^{\perp} \equiv F^{\#}(0)$ here, are $(0,1)$ and $(5,-2)$. Hence $\pi x \geq 0$ has the form $x_2 \geq 0$ and $5x_1 - 2x_2 \geq 0$. Since $d=2$ in our problem, both these define $d-2=0$ dimensional faces of $c\ell$ conv F, i.e., they define extreme points $(5,0)$ and $(2,5)$ respectively of $c\ell$ conv F. Hence case (ii) (b) of the converse applies. Further, $(5,0)$ is the intersection of adjacent facets $5s_1 + 3s_3 \geq 25$ and $-s_1 + s_3 \geq -5$ with $\dfrac{1}{25}(5,3) + \dfrac{1}{5}(-1,+1) =$ $(0, \dfrac{8}{25})$ which defines $\dfrac{8}{25} x_2 \geq 0$ or $x_2 \geq 0$. Similarly, $(2,5)$ is the intersection of adjacent facets $5s_1 + 3s_3 \geq 25$ and $-s_3 \geq -5$ with $\dfrac{1}{25}(5,3) + \dfrac{1}{5}(0,-1) =$ $(\dfrac{1}{5}, -\dfrac{2}{25})$ which defines $\dfrac{1}{5} x_1 - \dfrac{2}{25} x_2 \geq 0$ or $5x_1 - 2x_2 \geq 0$.

Implementation

Now, we know that if we obtain an extreme point $\pi(\neq 0)$ of $F^{\#} \cap L^{\perp}$, for $\pi_0 \neq 0$, then $\pi x \geq \pi_0$ defines a facet of $c\ell$ conv F if $0 \in$ aff F (or contains $c\ell$ conv F if $0 \notin$ aff F). Further, if $\pi(\neq 0)$ is an extreme direction of $F^{\#} \cap L^{\perp}$ with $\pi_0 = 0$, then $\pi x \geq 0$ either defines a facet of $c\ell$ conv F or again, contains $c\ell$ conv F. In any case, $\pi x \geq \pi_0$ is a valid inequality in the system defining $c\ell$ conv F. By virtue of Theorem 5.5, it is also sufficient to represent $c\ell$ conv F by the system of inequalities of the type $\pi x \geq \pi_0$ where π is either an extreme point of $F^{\#} \cap L^{\perp}$ if $\pi_0 \neq 0$ and is an extreme direction of $F^{\#} \cap L^{\perp}$ if $\pi_0 = 0$.

The basic problem at hand then, is to identify extreme points or extreme directions of $F^{\#} \cap L^{\perp}$, as appropriate. The concept utilized in accomplishing this is that if one minimizes a linear function g, say, over $F^{\#}$, and if the minimum value is finite, then one will have detected a lowest dimensional face of $F^{\#}$.

Hence, assume as before that $0 \notin c\ell$ conv F, that is, the current point (origin) is infeasible ·to the disjunction. Then the problem we wish to examine is

$$P1(g,\pi_0): \qquad \text{minimize} \quad \{g\pi: \pi \in F^{\#}(\pi_0)\}$$

Alternatively, using the characterization of $F^{\#}(\pi_0)$ given through Equation (5.17), we have

$$
\begin{aligned}
P1(g,\pi_0): \qquad &\text{minimize} \quad g\pi \\
&\text{subject to} \quad u^h D^h \leq \pi \\
&\qquad\qquad\quad u^h d^h \geq \pi_0 \\
&\qquad\qquad\quad u^h \geq 0, \ \pi \text{ unrestricted}
\end{aligned}
$$

where $u^h = (\theta^h, \sigma^h)$ and $D^h = \binom{D^h}{A}$.

The dual of this problem may be written as

$$
\begin{aligned}
P2(g,\pi_0): \qquad &\text{maximize } z = \sum_{h \in H^*} \pi_0 \xi_0^h \\
&\text{subject to} \quad D^h \xi^h \geq d^h \xi_0^h, \ h \in H^* \\
&\qquad\qquad\quad \sum_{h \in H^*} \xi^h = g \\[6pt]
&\qquad\qquad\quad \xi_0^h, \ \xi^h \geq 0, \ h \in H^*
\end{aligned}
$$

For our example problem, $P1(g,\pi_0)$, $P2(g,\pi_0)$ may be written as follows

P1(g,π_0):   minimize $g_1\pi_1 + g_2\pi_2$

subject to $-u_1^1 + u_2^1 + u_4^1 \leq \pi_1$

$u_1^1 - u_2^1 - u_3^1 - u_4^1 \leq \pi_2$

$-u_1^2 + u_2^2 \leq \pi_1$

$u_1^2 - u_2^2 - u_3^2 + u_4^2 \leq \pi_2$

$-5u_1^1 - 3u_2^1 - 5u_3^1 + 5u_4^1 \geq \pi_0$

$-5u_1^2 - 3u_2^2 - 5u_3^2 + 5u_4^2 \geq \pi_0$

$u_j^h \geq 0$, h=1,2, j=1,2,3,4

π_1, π_2 unrestricted

and

P2(g,π_0):   maximize $\xi_0^1 + \xi_0^2\ \pi_0$

subject to $-\xi_1^1 + \xi_2^1 \geq -5\xi_0^1$

$\xi_1^1 - \xi_2^1 \geq -3\xi_0^1$

$-\ \xi_2^1 \geq -5\xi_0^1$

$-\xi_2^1 \geq -5\xi_0^1$

$-\xi_1^2 + \xi_2^2 \geq -5\xi_0^2$

$\xi_1^2 - \xi_2^2 \geq -3\xi_0^2$

$-\ \xi_2^2 \geq -5\xi_0^2$

$\xi_2^2 \geq\ 5\xi_0^2$

$\xi_1^1 + \xi_1^2 = g_1$

$\xi_2^1 + \xi_2^2 = g_2$

$\xi_0^h,\ \xi_j^h \geq 0$, h=1,2; j=1,2

The task at hand then, is to find a characterization of g such that $P1(g,\pi_0)$ has a finite minimum, that is, $P2(g,\pi_0)$ is feasible with a finite maximum. Since $F^{\#} \neq \{\phi\}$, P1 is finite if and only if P2 is feasible. Such a characterization of g is given through the following theorem.

Theorem 5.6 (Characterization of g)

(i) If $g \in cl\ conv\ F$, $g \neq 0$, then for every $\lambda > 0$ such that $\lambda g \in cl\ conv\ F$ (and such a λ exists), we get

$P2(g,\pi_0)$ has a feasible solution $\bar{\xi} = (\bar{\xi}^h, \bar{\xi}_0^h)$, with $\sum_{h \in H^*} \bar{\xi}_0^h = 1/\lambda$

Conversely, if $\bar{\xi}$ is feasible to $P2(g,\pi_0)$ with $1/\lambda = \sum_{h \in H^*} \bar{\xi}_0^h$, then $g \in cl\ conv\ F$, $g \neq 0$, $\lambda g \in cl\ conv\ F$.

(ii) If $\pi_0 \neq 0$, then a solution $\bar{\xi}$ feasible to $P2(g,\pi_0)$ is also optimal if and only if the objective function value $\bar{z} = \pi_0(1/\bar{\lambda})$, where

$$\bar{\lambda} = \begin{cases} minimum\ \{\lambda:\ \lambda g \in cl\ conv\ F\}\ if\ \pi_0 > 0 \\ maximum\ \{\lambda:\ \lambda g \in cl\ conv\ F\}\ if\ \pi_0 < 0 \end{cases}$$

(iii) If $\pi_0 = 0$, any feasible solution $\bar{\xi}$ to $P2(g,0)$ is optimal with value $\bar{z} = 0$. However, $P1(g,0)$ has an optimal solution $(\bar{\pi}, \bar{u})$ with $\bar{\pi} g = 0$ and $\bar{\pi} \neq 0$ if and only if $g \in$ boundary of $cl\ conv\ F$.

Illustration

For the sake of illustration, let us consider part (ii) of the above theorem. Consider $\pi_0 = 1$ and $g = (1,1)$, say. Then,

$$\bar{\lambda} = minimum\ \{\lambda:\ \lambda(1,1) \in cl\ conv\ F\} = 25/8$$

Further, minimum $\{\pi_1 + \pi_2\}$ occurs at $(\frac{1}{5}, \frac{3}{25})$ with objective value
$\pi \in F^{\#}(1)$

$8/25 = \pi_0/\bar{\lambda} = 1/(25/8)$.

For part (iii) above, note that $g \in$ boundary $c\ell$ conv F and $\pi_0 = 0$ implies that $Pl(g,0)$: min $\{\lambda g: \pi \in F^\#(0)$ has an infinite number of alternative optimal solutions along some extreme direction of $F^\#(0) \cap L^\perp$. Note that $F^\#(0)$ is a poly-hedral cone with the vertex at the origin. Hence $(0,0)$ is also optimal and the optimal value is therefore always $\bar{\xi} = 0$.

Next, we give a characterization of solutions of $Pl(g,\pi_0)$ for a given g,π_0 such that these solutions contribute towards determining facets of $F^{\#\#}$ (and hence of $c\ell$ conv F) by detecting extreme points and extreme directions of $F^\#$. Such solutions are called <u>regular solutions</u>. Again, we consider cases $\pi_0 \neq 0$ and π_0 separately.

Theorem 5.7 (Regular Optimal Solutions)

(i) Suppose $\pi_0 \neq 0$. Then,

$\{\bar{\pi} \in F^\#$ is an extreme point of $F^\# \cap L^\perp\} \Leftrightarrow \{$(a) $\bar{\pi} \in L^\perp$ and (b) there exists a $p \in L^\perp$ such that $\bar{\pi} \in L^\perp$ is the unique point which minimizes πp on $F^\#\}$

Accordingly, if $(\bar{\pi},\bar{u})$ is optimal to $Pl(g,\pi_0)$ for some $g \in c\ell$ cone F, then

$\{\bar{\pi} \in$ vert $F^\# \cap L^\perp\} \Leftrightarrow \{$(1) $\bar{\pi} \in L^\perp$ and (2) there exists a $\gamma \in L^\perp$ such that if (π,u) solves $Pl(g + \gamma,\pi_0)$ and if $\pi \in L^\perp$ then this implies that $\pi = \bar{\pi}\}$

If $(\bar{\pi},\bar{u})$ satisfies these latter conditions (1) and (2), we call it a <u>regular solution</u>.

(ii) Suppose $\pi_0 = 0$. (Note: In this case, $F^\#$ is a cone and $F^\# \cap L^\perp$ is a pointed cone, that is, has an extreme point at the origin). Then,

$\{\bar{\pi} \in F^{\#}$ is an extreme direction of $F^{\#} \cap L^{\perp}\} <=> \{$(a) $\bar{\pi} \neq 0$, $\bar{\pi} \in L^{\perp}$ and

(b) there exists a

p $\in L^{\perp}$ such that up to

a positive multiplier,

$\bar{\pi} \in L^{\perp}$ is the unique

point which minimizes

πp on $F^{\#}\}$

Accordingly, if some $(\bar{\pi}, \bar{u})$ is optimal to $Pl(g, 0)$ for some $g \in$ boundary cℓ cone F (recall that this gives rise to extreme directions of $F^{\#}(0) \cap L^{\perp}$, then,

$\{\bar{\pi} \in$ dir $F^{\#} \cap L^{\perp}\} <=>$ (1) $\bar{\pi} \neq 0$, $\bar{\pi} \in L^{\perp}$ and (2) there exists a $\gamma \in L^{\perp}$

such that if (π, u) solves $Pl(g + \gamma, 0)$, and $\pi \neq 0$,

$\pi \in L^{\perp}$, then this implies that $\pi = \lambda\bar{\pi}$, $\lambda > 0$.

If $(\bar{\pi}, \bar{u})$ satisfies these latter conditions (1) and (2), we call it a

regular solution.

Interpretation

Consider Case (i). This simply says that for every extreme point $\bar{\pi}$ of $F^{\#} \cap L^{\perp}$, there exists a vector p which one may use in min $\{\pi p: \pi \in F^{\#}\}$ such that this extreme point is a unique minimizing point. Moreover, given an extreme point optimal solution $(\bar{\pi}, \bar{u})$ to $Pl(g, \pi_0)$, we can always perturb the objective function so that this extreme point is the unique optimal solution.

For example, in our illustrative problem, consider $\pi_0 = 1$. Then $(\frac{1}{5}, \frac{3}{25})$ is an extreme point of $F^{\#} \cap L^{\perp}$. Further, clearly,

(a) $(\frac{1}{5}, \frac{3}{25}) \in L^{\perp} \equiv R^2$, and

(b) taking p = (5,3), say, the problem min $5\pi_1 + 3\pi_2: \pi \in F^{\#}(1)$ has $(\frac{1}{5}, \frac{3}{25})$ as its unique optimal solution.

Similar remarks hold for Case (ii). Here, the uniqueness of the optimal solution is upto a positive multiplier since we always encounter alternative optimal minimizing points along extreme directions in this case.

An important and relevant result which ties in the above statements is given next.

Lemma 5.3

If $Pl(g,\pi_0)$ has an optimal solution, then it has a regular optimal solution.

We will now characterize facets of $F^{\#\#}$ (and hence of $cl \ conv \ F$) in terms of regular optimal solutions to $Pl(g,\pi_0)$.

Theorem 5.8 (Characterization of facets of $F^{\#\#}$, and hence facets of $cl \ conv \ F$, in terms of regular optimal solutions)

(i) Suppose $\pi_0 \neq 0$. Let $g \ \epsilon \ cl \ cone \ F$, $g \neq 0$ and let

$$\bar{\lambda} = \begin{cases} \text{minimum } \{\lambda: \lambda g \ \epsilon \ cl \ conv \ F\} \text{ if } \pi_0 > 0 \\ \text{maximum } \{\lambda: \lambda g \ \epsilon \ cl \ conv \ F\} \text{ if } \pi_0 < 0 \end{cases}$$

Then, $\pi x > \pi_0$, $(\pi_0 \neq 0)$, $\pi \ \epsilon \ \ell h \ F$ is a facet of $F^{\#\#}$ containing the point $\bar{\lambda}g$, if and only if $\pi = \bar{\pi}$ for some regular optimal solution $(\bar{\pi},\bar{u})$ to $Pl(g,\pi_0)$.

(ii) Suppose $\pi_0 = 0$. Let $g \ \epsilon \ boundary \ cl \ conv \ F$, $g \neq 0$. Then, $\pi \ x \geq 0$, $\pi \ \epsilon \ \ell h \ F$ is a facet of $F^{\#\#}$ containing the point g if and only if $\pi = \lambda\bar{\pi}$ for some $\lambda > 0$ and some regular optimal solution $(\bar{\pi},\bar{u})$ to $Pl(g,0)$.

Illustration

Consider Case (i) above, and let $\pi_0 = 1$. Consider the facet $\frac{5s_1}{25} + \frac{3s_3}{25} \geq 1$ of $F^{\#\#}$ and recall that for our problem $\ell h \ F = R^2$. Now, since this is also a facet of $cl \ conv \ F$, hence for any $g \ \epsilon \ cl \ conv \ F$, we can find the appropriate $\bar{\lambda}$ such that $\bar{\lambda} g$ lies on this facet. Moreover, $\pi = (\frac{1}{5},\frac{3}{25})$ here is a regular optimal solution to $Pl(g,1)$ for any $g \ \epsilon \ cl \ conv \ F$.

Next, consider Case (ii). Here, the vector $g = (2,5)$, say, belongs to the boundary of $cl \ conv \ F$, $g \neq 0$. Then, $5s_1 - 2s_3 \geq 0$ is a facet of $F^{\#\#}$ (0) containing $(2,5)$. Moreover, taking $\lambda = 1$, we see that $\pi = (5,-2) = \bar{\pi}$ solves the problem to minimize $2\pi_1 + 5\pi_2$ subject to $\pi \ \epsilon \ P^{\#}(0)$. In fact, any point along the ray satisfying $2\pi_1 + 5\pi_2 = 0$ solves this latter problem. Hence, $\pi = \lambda\bar{\pi}$, $\lambda = 1$ for some regular optimal solution $(\bar{\pi},\bar{u})$ to $Pl(g,0)$.

<u>Summary and Notes</u>: Using Theorem 5.7 and 5.8 above, all the facets of $F^{\#\#}$ may be obtained by solving $Pl(g,\pi_0)$ (or its dual) for various vectors $g \in c\ell$ conv F. Further, from Theorem 5.5, each such facet is, or yields in conjunction with some other facets of $F^{\#\#}$, a facet of $c\ell$ conv F. The following points are worth noting:

(i) If $\pi_0 \neq 0$, $\bar{\lambda}$ is as defined in Theorem 5.8(i), and if $\bar{\lambda} g$ is the convex combination of k extreme points and extreme directions of $c\ell$ conv F $(1 \leq k \leq n)$ then this implies that each of these vertices and extreme direction vectors are contained in each facet of $c\ell$ conv F that contains $\bar{\lambda} g$. Moreover, each such facet can be obtained by solving $Pl(g,\pi_0)$ (or its dual).

For example, refer to Figure 5.4 below. $\bar{\lambda} g$ is the convex combination of extreme points $(2,5)$ and $(5,0)$ of $c\ell$ conv F. Only one such facet contains these extreme points and also $\bar{\lambda} g$, and this facet is generated by solving $Pl(g,\pi_0)$ as demonstrated in the illustrative example solved below in Section 5.5.

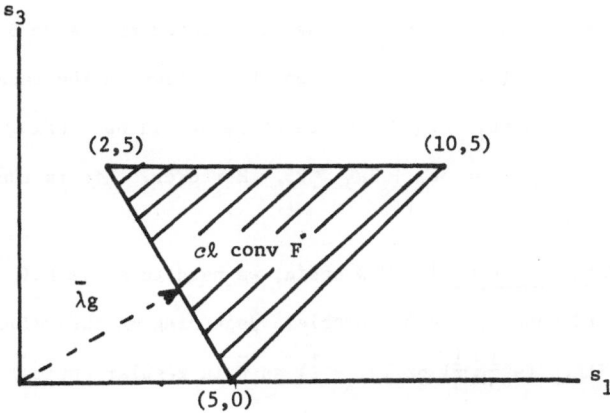

Figure 5.4. Representation of $\bar{\lambda}g$

(ii) Analogous remarks hold for the case $\pi_0 = 0$ as in (i) above.

(iii) If g corresponds to an extreme point of $c\ell$ conv F, then solving $Pl(g,\pi_0)$ with $\pi_0 = 1, -1$ and 0 gives all the facets of $c\ell$ conv F containing the vertex g. This is a special case of remark (i) above. Furthermore, for a given π_0, the associated facets all correspond to alternative regular optimal solutions of $Pl(g,\pi_0)$. Thus, if one facet containing g is found, the others are easily obtained therefrom.

Specialized schemes for solving $Pl(g,\pi_0)$ and $P2(g,\pi_0)$ may be devised and are in fact available. However, since this leads away from the motivation of our present disucssion, we avoid these schemes here. We next illustrate in detail the method of determining facets of $c\ell$ conv F for our example problem.

5.5 Illustrative Example

Consider the example problem we introduced in Section 5.1. We will first obtain facets of the "≥ 1 type", then of the "≤ -1 type" and finally of the "≥ 0 type for $c\ell$ conv F.

1. "≥ 1 type" facets ($\pi_0=1$). Here, we are interested in solving $Pl(g,1)$ for various vectors g ε $c\ell$ cone F. If $(\bar{\pi},\bar{u})$ is obtained as a regular optimal solution, then we put $\pi = \bar{\pi}$ and derive $\pi x \geq 1$ as a facet of $F^{\#\#}$ and hence as a facet of $c\ell$ conv F.

   We note here, that for any g ε $c\ell$ cone F, we get $\bar{\pi} = (\frac{2}{5},\frac{3}{25})$ in the regular optimal solution. Also, note that as illustrated in Figure 5.4, defining $\bar{\lambda}$ as in Theorem 5.8(i), for g ε $c\ell$ cone F, $\bar{\lambda}$g lies on the facet $5s_1 + 3s_3 \geq 25$ of $c\ell$ conv F. Indeed, this is the facet we obtain here using $\pi = \bar{\pi}$ in $\pi x \geq 1$, that is $\frac{s_1}{5} + \frac{3s_3}{25} \geq 1$ or $5s_1 + 3s_3 \geq 25$. Moreover, this is the only "≥ 1 type" facet $c\ell$ conv F.

2. "≥ -1 type" facets ($\pi_0=-1$). Now again, we need to solve $Pl(g,-1)$ for various vectors g ε $c\ell$ cone F. In our problem, depending on the values of g selected, we obtain either $(-\frac{1}{5}, \frac{1}{5})$ or $(0, -\frac{1}{5})$ as $\bar{\pi}$ in regular optimal solutions. Thus,

$\dfrac{-s_1}{5} + \dfrac{s_3}{5} \geq -1$ and $\dfrac{-s_3}{5} \geq -1$ are facets of $F^{\#\#}$ (-1) and hence of $c\ell$ conv F.

Thus, the only "≥ -1 type" facets of $c\ell$ conv F are $s_1 - s_3 \leq 5$ and $s_3 \leq 5$.

3. "≥ 0 type" facets $(\pi_0 = 0)$. Now, we need to solve $Pl(g,0)$ for vectors $g \in$ boundary $c\ell$ cone F, $g \neq 0$. Thus, we may either select

$$g = \lambda(2,5), \ \lambda > 0$$

or

$$g = \lambda(5,0), \ \lambda > 0$$

The first choice of g yields

$$\bar{\pi} = \beta(5,-2), \ \beta > 0 \text{ in the regular optimal solution to } Pl(g,0)$$

and the second choice yields

$$\bar{\pi} = \beta(0,1), \ \beta > 0$$

This gives us $5s_1 - 2s_3 \geq 0$ and $s_3 \geq 0$ as facets of $F^{\#\#}(0)$. (Note that $(5,-2)$ and $(0,1)$ are extreme directions of $F^{\#}(0) \cap L^{\perp} = F^{\#}(0)$ here, since $L^{\perp} = R^2$). Now, according to case (ii) of Theorem 5.5, consider the facet $5s_1 - 2s_3 \geq 0$. (Similar remarks hold for the facet $s_3 \geq 0$). Either this is a facet of $c\ell$ conv F or it defines a 0-dimensional (d-2 dimensional in general) intersection of facets of $c\ell$ conv F. As observed earlier, in either case, we can use these inequalities as defining half-spaces for $c\ell$ conv F without any inhibitions. At most, as seen in Figure 5.5 below, we will have introduced degeneracy in the problem. Thus, the set $c\ell$ conv F obtained above may be defined by the inequalities.

$$5s_1 + 3s_3 \geq 25$$

$$s_1 - s_3 \leq 5$$

$$s_3 \leq 5$$

$$5s_1 - 2s_3 \geq 0$$

$$s_3 \geq 0$$

These inequalities may now replace the disjunctive statement in Problem DP. The set $c\ell$ conv F is depicted in Figure 5.5 below.

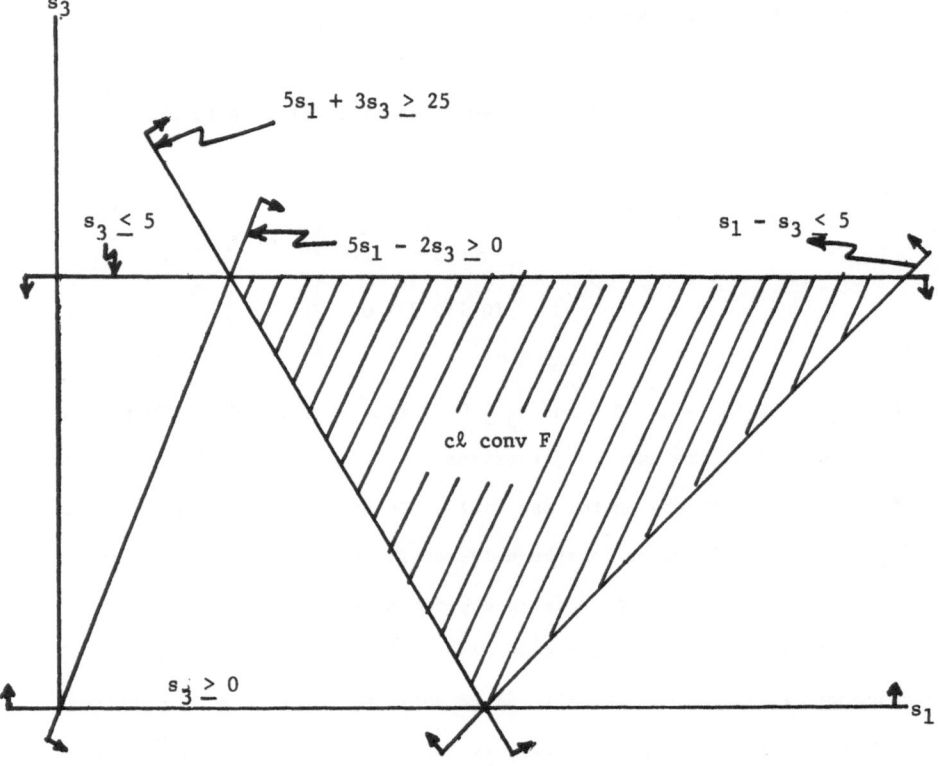

Figure 5.5. Inequalities Defining $c\ell$ conv F

5.6 Facial Disjunctive Programs

As may be apparent from our foregoing discussion, if $|H|$ is large, then the solution of Problem DP through the generation of facets of $c\ell$ conv F is prohibitive. For example most mixed integer linear programs would be intractable by this technique due to the size and complexity of Problems $P1(g,\pi_0)$ and $P2(g,\pi_0)$. However, this technique is attractive for small $|H|$.

Hence, for a large $|H|$ there is a need to relax some disjunction and in some manner, use facets obtained for some enforced disjunctions to generate facets for other disjunctions when they are also simultaneously enforced. It turns out that such a procedure is possible for special disjunctive programs called <u>facial disjunctive problems</u>.

Before we discuss this, let us consider Problem DP. This problem has been stated in the so called <u>disjunctive normal form</u>. There is another way of writing Problem DP, which we will find more convenient in the present context. Suppose that for each $h \in H$, $A^h x \geq b^h$ has $|Q_h = Q|$ inequalities, where Q is the index set of these inequalities. Then, we may construct $|Q|$ sets H_j, $j \in Q$ such that for each $h \in H$, exactly one of the inequalities in $A^h x \geq b^h$ is placed distinctly in each of the sets H_j, $j \in Q$. Thus, H is the cartesian product of the H_j, $j \in Q$, that is,

$$H = \prod_{j \in Q} H_j \qquad (5.24)$$

Problem DP may now be stated in the so-called <u>conjunctive normal form</u>

$$\begin{aligned}
\text{minimize} \quad & f(x) = c^t x \\
\text{subject to} \quad & x \in F_0 = \{x: Dx \geq d, \ x \geq 0\}
\end{aligned}$$

$$\bigwedge_{j \in Q} [\bigvee_{i \in H_j} a_i^h x \geq b_i^h]$$

where,

$$\{x: A^h x \geq b^h, \; x \geq 0\} \equiv \{x: a_i^h x \geq b_i^h, \; i \varepsilon Q, \; x \geq 0\}, \; h \varepsilon H \qquad (5.25)$$

and F_0 is as defined in Equation (5.3). Now consider the set F defined in Equation (5.2) and let us denote it by F^Q. Accordingly, for a set $T \subset Q$, let us denote the corresponding set of feasible points to the (relaxed) disjunction as F^T. Thus,

$$T \subset Q \Rightarrow F^T \supseteq F^Q \qquad (5.26)$$

The set $c\ell \; \text{conv} \; F^T$ will be called a partial convex hull of F for $T \subset Q$. Now, suppose we use the disjunctions in some set $T \subset Q$ alone and as before generate all facets of F^T. Further, suppose that we now replace the disjunctions in T with these facets in the original problem. Does the solution of the resulting problem satisfy the disjunctions in T? The answer is yes for special problems called facial disjunctive programs described below.

Consider the following definitions.

Definition 5.1

Let F_0 be a convex set. A subset F of F_0 (possibly empty) is called a face of F_0 if there exists a supporting hyperplane of F_0 whose intersection with F_0 defines F.

Definition 5.2

A disjunction $\bigvee\limits_{i \varepsilon H_j} \{a_i^h x \geq b_i^h\}$ is called facial with respect to F_0, if

$$F_i^h = F_0 \cap \{x \varepsilon R^n: a_i^h x \geq b_i^h\} \qquad (5.27)$$

is a face of F_0 for each $i \varepsilon H_j$. (Note that a face may be an extreme point, an edge, ..., a facet or the entire set). A disjunctive program is said to be a facial disjunctive program if F_1 is a face of F_0 for each $i \varepsilon H_j$ and for each $j \varepsilon Q$.

<u>EXAMPLE</u>. In our problem, we have,

$$F_1^1 = F_0 \cap \{(s_1,s_3): s_3 \geq 5\}.$$

and

$$F_1^2 = F_0 \cap \{(s_1,s_3): s_1 - s_3 \geq 5\}.$$

These are depicted as F_1 and F_2 respectively in Figure 5.1. As one may see in that figure, F_1^1 and F_1^2 are faces (facets in this case) of F_0. Hence, our problem is <u>facial</u>.

<u>Necessary and Sufficient Condition for $a_i^h x \geq b_i^h$ to be a Face of F_0: Theorem 5.9.</u>
Let F_i^h be as defined in Equation (5.27). If there exist $(\mu,\nu) \in R^m \times R^n$ satisfying

$$\left\{ \begin{array}{l} \mu(-D) + \nu(-I) = a_i^h \\[2ex] \mu(-d) \qquad\quad = b_i^h \\[2ex] \qquad (\mu,\nu) \geq 0 \end{array} \right\} \qquad (5.28)$$

then F_i^h is a face of F_0, namely,

$$F_i^h = \{x \epsilon F_0: a_i^h x = b_i^h\} = \{x \epsilon F_0: D_i x = d_i \text{ for each } i \epsilon M^+,$$

$$x_j = 0 \text{ for each } j \epsilon N^+\} \qquad (5.29)$$

where D_i is the ith row of D and

$$M^+ = \{i \epsilon M: \mu_i > 0\}, \ N^+ = \{j \epsilon N: \nu_j > 0\} \qquad (5.30)$$

Conversely, if F_j^h is a face of F_0, and $F_0 \neq F_i^h \neq \{\phi\}$, then there exist $(\mu,\nu) \epsilon R^m \times R^n$

satisfying the property (5.28).

Henceforth, we will assume that F_0 is bounded (we may regularize it if necessary) and that DP is facial. This implies that F_h of Equation (5.1) is a polytope for each $h \in H$, that is conv F^S ($\equiv$conv F) is a polytope.

Consequence of the Facial Property:

Theorem 5.10

If DP is facial and F_0 is bounded, then

$$\{\text{Extreme points of conv } F^Q\} = \bigcup_{h \in H} \{\text{extreme points of } F_h\}$$

$$\subseteq \{\text{extreme points of } F_0\}$$

For example, in our problem, referring to Figure 5.1, we have,

$\{\text{extreme points of conv } F\} = \{(2,5),(10,5),(5,0)\}$

$\{\text{extreme points of } F_1\} \quad = \{(2,5),(10,5)\}$

$\{\text{extreme points of } F_2\} \quad = \{(10,5),(5,0)\}$

$\{\text{extreme points of } F_0\} \quad = \{(0,0),(0,3),(2,5),(10,5),(5,0)\}$

Practically speaking, the most important consequence of the facial property is that conv F^Q may be obtained in as many steps $(|Q|)$, as there are disjunctions in the conjunctive normal form, by applying the disjunctions one at a time alone. Now, when DP is facial and F_0 is bounded, it turns out that if $T \subset Q$, $i \in H_j$ and $j \in Q-T$, then,

$$F_i^h \cap \text{ conv } F^T = \text{conv}[F_i^h \cap F^T] \tag{5.31}$$

In other words, having conv F^T, for some $T \subset Q$, we select an F_i^h not yet considered and compute conv$[F_i^h \cap F^T]$ simply as $F_i^h \cap$ conv F^T. This leads to another important result.

Theorem 5.11

Assume that DP is facial and that F_0 is bounded. Then, for any $T \subset Q$,

$$\text{conv}[F^{Q-T} \cap \text{conv } F^T] = \text{conv } [F^Q]$$

This main result is used to compute $\text{conv}[F^Q]$ in the following manner. Let $Q=\{j_1,j_2,\ldots,j_q\}$ where $q = |Q|$. Then as a corollary to the above theorem, we have,

$$\text{conv } F^Q = \text{conv}[F^{\{j_q\}} \cap \text{conv}\{\ldots \cap \text{conv}(F^{\{j_2\}} \cap \text{conv } F^{\{j_1\}})\ldots)\}] \qquad (5.32)$$

We may now apply (5.31) to the decomposition (5.32) in order to compute $\text{conv } F^Q$ in $q = |Q|$ steps.

We terminate a brief discussion of facial disjunctive programs at this point. Later, in Chapter VII, we will return to facial disjunctive programs as a special case of Problem DP and will present two finitely convergent algorithms to solve such problems. One of these procedures is based on Theorem 5.10 whereas the other is based on Theorem 5.11. Both of these procedures solve Problem DP by generating facets of $c\ell$ $\text{conv } F^Q$ as and when needed till either an optimal solution is obtained or all the facets of $c\ell$ $\text{conv } F^Q$ have been generated, whence the problem is necessarily solved.

Thus far, we have addressed the question of generating deep disjunctive cuts. In the next chapter, we will examine some of the cutting planes available in the literature and identify them as basically disjunctive cutting planes by putting them in the general format of the latter type of cuts.

5.7   Notes and References

This chapter is heavily based on the results of Balas in [5]. If the facets of the closure of the convex hull of feasible points are known, clearly the problem of solving disjunctive programs is trivial. Balas' study takes an important step in characterizing them. Furthermore, for a special important case of disjunctive programs, the study opens up the possibility of generating the facets sequentially.

Chapter VI

DERIVATION AND IMPROVEMENT OF SOME EXISTING CUTS THROUGH
DISJUNCTIVE PRINCIPLES

6.1 Introduction

In discussing the basic disjunctive cut principle, we indicated that it
subsumes all other cut generation principles. In this chapter, we will demonstrate
this to a certain extent by actually deriving some existing cutting planes as
disjunctive cuts. In the process, it will be seen that the disjunctive principles
may be used to actually improve upon three cuts. In fact, for the first type of
cut we discuss below, we will utilize the concepts of Chapter IV to obtain an
improved version of the existing cut.

6.2 Gomory's Mixed Integer Cuts

Consider a mixed integer program where only certain variables are constrained
to be integral. Suppose we have a simplex tableau representation of a basic
feasible solution to the corresponding problem with integrality relaxed. Further,
assume that this solution does not satisfy the integrality constraints. In
particular, let us identify a basic, integer-constrained variable x_i whose current
value, a_{i0}, is non-integral. Let us write the representation of x_i in terms of
the non-basic variables t_j, $j \varepsilon J$ in the tableau representing the solution at hand
as follows

$$x_i = a_{i0} + \sum_{j \varepsilon J} a_{ij}(-t_j) \qquad (6.1)$$

Now, partition J as

$$J = J_1 \cup J_2 \qquad (6.2)$$

where

$$J_1 = \{j \epsilon J: t_j \text{ is integer-constrained}\} \qquad (6.3)$$

and

$$J_2 = J - J_1 \qquad (6.4)$$

Further, denoting the largest integer less than or equal to a given real number μ as $[\mu]$ and denoting the smallest integer greater than or equal to a given real number μ as $\langle\mu\rangle$, we may write

$$a_{i0} = [a_{i0}] + f_{i0}, \ f_{i0} > 0 \qquad (6.5)$$

$$a_{ij} = [a_{ij}] + f_{ij}, \ \text{for } j \epsilon J \qquad (6.6)$$

Substituting (6.5), (6.6) into (6.1), we obtain

$$x_i = [a_{i0}] + f_{i0} + \sum_{j \epsilon J_1} ([a_{ij}] + f_{ij})(-t_j) + \sum_{j \epsilon J_2} a_{ij}(-t_j)$$

or

$$x_i - [a_{i0}] - \sum_{j \epsilon J_1} [a_{ij}](-t_j) = f_{i0} + \sum_{j \epsilon J_1} f_{ij}(-t_j) + \sum_{j \epsilon J_2} a_{ij}(-t_j) \qquad (6.7)$$

Now, let us introduce a new set of parameters ϕ_{ij}, $j \epsilon J \cup \{0\}$ defined as follows

$$\phi_{i0} = f_{i0}$$

$$\phi_{ij} = a_{ij} \ \text{for } j \epsilon J_2$$

$$\phi_{ij} = \begin{cases} f_{ij} & \text{if } f_{ij} \leq f_{i0} \\ \\ f_{ij}-1 & \text{if } f_{ij} > f_{i0} \end{cases} \qquad \text{for } j \epsilon J_1 \qquad (6.8)$$

These parameters are merely notational expedients. Substituting (6.8) into (6.7), we obtain

$$x_i - [a_{i0}] - \sum_{j \in J_1} [a_{ij}](-t_j) - \sum_{\substack{j \in J_1 \\ f_{ij} > f_{i0}}} (-t_j) = \phi_{i0} + \sum_{j \in J} \phi_{ij}(-t_j) \qquad (6.9)$$

Finally, denote

$$y_i = \phi_{i0} + \sum_{j \in J} \phi_{ij}(-t_j) \qquad (6.10)$$

Observe in Equation (6.9) that the left hand side is necessarily integral and hence, so is the right hand side. That is, y_i of Equation (6.10) should necessarily be integral. In particular, the following disjunction must hold

$$\{y_i \leq 0\} \ \lor \ \{y_i \geq 1\} \qquad (6.11)$$

that is,

$$\{\sum_{j \in J} \phi_{ij} t_j \geq \phi_{i0}\} \ \lor \ \{\sum_{j \in J} -\phi_{ij} t_j \geq 1 - \phi_{i0}\}$$

Noting from (6.5), (6.8) that $0 < \phi_{i0} = \dot{f}_{i0} < 1$ whenever x_i is fractional, we may use the basic disjunctive cut principle along with the deep cut notion of Section 3.2 to first write the above disjunction as

$$\{\sum_{j \in J} \frac{\phi_{ij}}{\phi_{i0}} t_j \geq 1\} \ \lor \ \{\sum_{j \in J} \frac{-\phi_{ij}}{1-\phi_{i0}} t_j \geq 1\}$$

and then derive the cut

$$\sum_{j \in J} \max\{\frac{\phi_{ij}}{\phi_{i0}}, \frac{-\phi_{ij}}{1-\phi_{i0}}\} \ t_j \geq 1 \qquad (6.12)$$

Note in Equation (6.12), for each $j \varepsilon J$, the cut coefficient is determined by the nonnegative element of the pair. Thus, if we define

$$J_k^+ = \{j \varepsilon J_k : \phi_{ij} \geq 0\}, \ J_k^- = \{j \varepsilon J_k : \phi_{ij} < 0\} \text{ for } k=1,2 \tag{6.13}$$

we may write (6.12) as

$$\sum_{j \varepsilon J_1^+} \frac{\phi_{ij}}{\phi_{i0}} t_j + \sum_{j \varepsilon J_1^-} \frac{-\phi_{ij}}{1-\phi_{i0}} t_j + \sum_{j \varepsilon J_2^+} \frac{\phi_{ij}}{\phi_{i0}} t_j + \sum_{j \varepsilon J_2^-} \frac{-\phi_{ij}}{1-\phi_{i0}} t_j \geq 1$$

Finally, substituting (6.8) into this cut, we obtain Gomory's mixed integer cut as

$$\sum_{j \varepsilon J_1^+} \frac{f_{ij}}{f_{i0}} t_j + \sum_{j \varepsilon J_1^-} \frac{1-f_{ij}}{1-f_{i0}} t_j + \sum_{j \varepsilon J_2^+} \frac{a_{ij}}{f_{i0}} t_j + \sum_{j \varepsilon J_2^-} \frac{-a_{ij}}{1-f_{i0}} t_j \geq 1 \tag{6.14}$$

Now, let us consider improving this cut. The concept of the strategy we employ is basically that of Chapter IV. More specifically, we reformulate the disjunction (6.11) to incorporate additional constraints as follows

$$\begin{Bmatrix} y_i \leq 0 \\ x_h \geq 0, \ h \varepsilon I \end{Bmatrix} \vee \begin{Bmatrix} y_i \geq 1 \\ x_h \geq 0, \ h \varepsilon I \end{Bmatrix} \tag{6.15}$$

where I denotes the set of basic variables. Hence, letting

$$x_h = a_{h0} + \sum_{j \varepsilon J} a_{hj}(-t_j) \quad \text{for } h \varepsilon I \tag{6.16}$$

we may rewrite (6.15) by using (6.10), (6.16) as

$$\left\{\begin{array}{l} \sum_{j \in J} \phi_{ij} t_j \geq \phi_{i0} \\[2ex] \sum_{j \in J} (-a_{hj}) t_j \geq -a_{h0}, \ h\epsilon I \end{array}\right\} \vee \left\{\begin{array}{l} \sum_{j \in J} (-\phi_{ij}) t_j \geq 1 - \phi_{i0} \\[2ex] \sum_{j \in J} (-a_{hj}) t_j \geq -a_{h0}, \ h\epsilon I \end{array}\right\} \qquad (6.17)$$

Letting λ_0^1, λ_h^1, hϵI be the nonnegative multipliers for the first set of constraints in (6.17) and λ_0^2, λ_h^2, hϵI those for the second set, we may write the appropriate surrogate constraints as

$$\sum_{j \in J} (\lambda_0^1 \phi_{ij} - \sum_{h \epsilon I} \lambda_h^1 a_{hj}) t_j \geq (\lambda_0^1 \phi_{i0} - \sum_{h \epsilon I} \lambda_h^1 a_{h0})$$

and

$$\sum_{j \in J} (-\lambda_0^2 \phi_{ij} - \sum_{h \epsilon I} \lambda_h^2 a_{hj}) t_j \geq \lambda_0^2 (1 - \phi_{i0}) - \sum_{h \epsilon I} \lambda_h^2 a_{h0}$$

Using the concepts of Chapter III, the disjunctive cut we derive is

$$\sum_{j \in J} \max\{\lambda_0^1 \phi_{ij} - \sum_{h \epsilon I} \lambda_h^1 a_{hj}, \ -\lambda_0^2 \phi_{ij} - \sum_{h \epsilon I} \lambda_h^2 a_{hj}\} t_j \geq 1 \qquad (6.18)$$

where

$$\lambda_0^1 \phi_{i0} - \sum_{h \epsilon I} \lambda_h^1 a_{h0} = 1$$

$$\lambda_0^2 (1 - \phi_{i0}) - \sum_{h \epsilon I} \lambda_h^2 a_{h0} = 1$$

$$\lambda_0^1, \ \lambda_h^1, \ \lambda_0^2, \ \lambda_h^2 \geq 0 \qquad h\epsilon I \qquad (6.19)$$

Clearly, (6.18) can be made to uniformly dominate Gomory's mixed integer cut (6.12) (or(6.14)) since the latter is obtained from the former by selecting $\lambda_0^1 = 1/\phi_{i0}$, $\lambda_0^2 = 1/(1-\phi_{i0})$ and $\lambda_h^k = 0$, k=1,2, hϵI. Again, for the appropriate selection of parameters in (6.18), (6.19), one may resort to the concepts of Chapter III. Alternatively, one may handle the constraints $x_h \geq 0$, $h\varepsilon I$ of Equation (6.15) in a manner similar to that recommend in Chapter IV.

6.3 Convexity or Intersection Cuts with Positive Edge Extensions

In this section, we will discuss the general setting through which convexity or intersection cuts are derived with the purpose of demonstrating how the disjunctive cut principle is capable of generating such cuts. Hence, consider a convex set

$$C = \{x: a^h x \leq b^h, \ h\varepsilon H\} \tag{6.20}$$

defined by certain hyperplanes $a^h x \leq b^h$, $h\varepsilon H$, where $a^h = (a_1^h, \ldots, a_n^h)$. Further, let $b^h > 0$, $h\varepsilon H$ and suppose that we are currently located at the origin. In this setting, let us assume that there exists a subset S of the nonnegative orthant of R^n which contains points of interest to us. Suppose we have identified a set C which contains the origin, but no point of S, i.e., $S \cap C = \{\phi\}$. The point we are currently located at, viz., the origin, is not of (further) interest to us. Our intention now is to use the set C to generate a cut which deletes the origin but no point of S.

Accordingly, let us identify the n half lines

$$\xi^j = \{x: x = \lambda e_j, \ \lambda \geq 0\} \quad j=1,\ldots,n \tag{6.21}$$

where e_j is the jth unit vector. These half lines are defined by the coordinate axes incident at the origin. Let us now proceed along each of the half lines (6.21) in turn and compute the maximum distance $\bar{\lambda}_j$ we can traverse along this direction and still remain within the set C. In other words,

$$\bar{\lambda}_j = \sup\{\lambda \geq 0: (a^h)(\lambda e_j) \leq b^h, \ h\varepsilon H\} \quad \text{for } j=1,\ldots,n \tag{6.22}$$

Then, it can be shown that a valid convexity cut is given by

$$\sum_{j=1}^{n} (1/\bar{\lambda}_j)x_j \geq 1 \qquad (6.23)$$

Typically, one works with the simplex tableau representation of the current point, whence, x_j, $j=1,\ldots,n$ are the nonbasic variables. Accordingly, C is defined in terms of nonbasic variables and at the current point, $x_j=0$, $j=1,\ldots,n$.

Now, let us apply the disjunctive principle to this situation. Observe that since $S \cap C = \{\phi\}$, we are only interested in those points $x \geq 0$ which violate at least one of the inequalities defining C. Thus, we may stipulate that at least one of the systems

$$a^h x \geq b^h \qquad h\epsilon H$$

$$x \geq 0$$

holds. Using Theorem 3.1, a suitable cut which one may derive through Equation (3.16) is given by

$$\sum_{j=1}^{n} \{\sup_{h\epsilon H}(a_j^h/b^h)\}x_j \geq 1 \qquad (6.24)$$

Note that one may have preferably chosen the cut given by Equation (3.19) instead. However, we will work with the above cut to preserve simplicity as well as to derive certain known cuts in the literature. Returning to Equation (5.22), we observe that

$$\bar{\lambda}_j = \begin{cases} \inf_{h\epsilon H}\{b^h/a_j^h: a_j^h > 0\} \\ \\ \infty \text{ if } a_j^h \leq 0 \text{ for each } h\epsilon H \end{cases} \qquad j=1,\ldots,n$$

Hence, the cut (6.23) has coefficients

$$
1/\bar{\lambda}_j = \begin{cases} \sup_{h \in H}\{a_j^h/b^h : a_j^h > 0\} \text{ if at least one } a_j^h > 0 \\[2mm] 0 \text{ if all } a_j^h \leq 0 \end{cases} \qquad j=1,\ldots,n \qquad (6.25)
$$

Comparing (6.24) and (6.25), we note that if $\xi^j \not\subset C$ for any $j \varepsilon \{1,\ldots,n\}$, then at least one $a_j^h > 0$, $h \varepsilon H$ for each $j \varepsilon \{1,\ldots,n\}$. In this case, (6.24) and (6.25) are identical. However, if $\xi^j \subset C$ for at least one $j \varepsilon \{1,\ldots,n\}$, then the corresponding coefficient $1/\bar{\lambda}_j$ is zero for the cut (6.23) but may be negative for the cut (6.24). Hence, (6.24) uniformly dominates (6.23), and may strictly dominate it. This latter case is depicted in Figure 6.1 below.

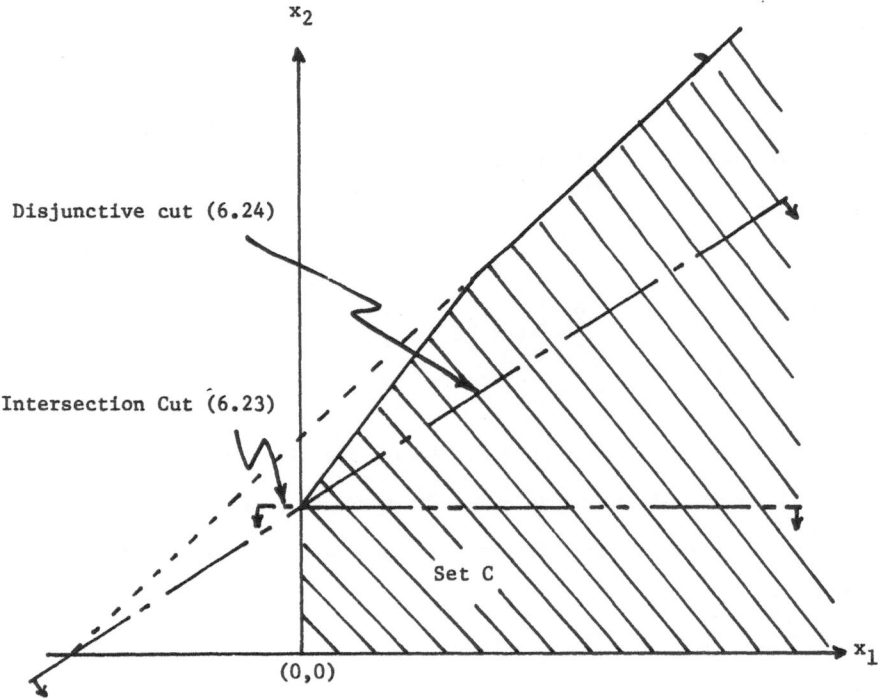

Figure 6.1.   Illustration of Disjunctive and Intersection Cuts

Several existing cuts are subsumed under the category just described.
Notice that we have not required the set H to be of finite cardinality. Thus,
C need not be polyhedral. Therefore, hypercylindrical or spherical intersection
cuts are also recovered under the discussed framework. In this situation, a
hypercylinder or sphere containing the origin but no integer valued points
(integral in terms of original variables) may be defined. Accordingly, the constraints in C represent tangential hyperplanes to the hypercylinder or sphere.
If one uses a stronger condition and requires each constraint of C to simply
correspond to a tangent to the sphere at some integer point, then one recovers
octahedral cuts. Similarly, to obtain diamond cuts, one may use the condition
that at least $|I|$ of the $2|I|$ constraints $x_i \leq 0$, $x_i \geq 1$, $i \epsilon I$ must hold. This
latter cut may be further strengthened by replacing x_i by y_i, $i \epsilon I$ as defined in
Equation (6.10).

Again, various polar cutting planes may be recovered from the above discussion by letting C be an appropriate reverse polar set. In this connection,
the reader may note that the negative edge extension cut and the reverse polar cut
would be identical to the disjunctive cut derived above.

We will now proceed to discuss one such special case in the next section.
We will show how the convexity cuts generated in this case are subsumed under the
disjunctive cut principle and how these cuts may be further strengthened.

6.4 Reverse Outer Polar Cuts for Zero-One Programming

Consider a program in (x,y) of the form

$$
\begin{aligned}
\text{minimize} \quad & cx + dy \\
\text{subject to} \quad & A(x,y) = b \\
& x \epsilon S \\
& x,y \geq 0
\end{aligned}
$$

Here, S is some set of points of interest to us. Also, currently, suppose we
have a basic solution $(\bar{x}, \bar{y})$ to the linear program with the constraint $x \epsilon S$ relaxed.

Further, assume that $\bar{x} \notin S$. Now, let us define the set C. For this purpose, let us say that we can identify some bilinear function

$$f(x,z): R^n \times R^n \to R \qquad (6.26)$$

and a scalar k which are such that if we let C be the level set

$$C = L(k) = \{x: f(x,x) \leq k\} \qquad (6.27)$$

then C contains $\bar{x}$, but int C contains no point in $S \cap X$, where,

$$X = \{x \geq 0: A(x,y) = b \text{ for some } y \geq 0\} \qquad (6.28)$$

One may now polarize the function $f(.,.)$ by replacing its argument (x,x) by (x,z) in (6.27) and thereby define a <u>reverse polar set</u>, alternatively called the <u>scaled generalized reverse polar</u> of X, with scalar k, as

$$X^o(k) = \{z \epsilon R^n: f(x,z) \leq k, \text{ for each } x \epsilon X\}$$

$$= \{z \epsilon R^n: f(x,z) \leq k, \text{ for each } x \epsilon \text{ vert } X\} \qquad (6.29)$$

where we have assume that X is bounded and that vert X represents the set of extreme points X. Note that although $f(.,.)$ or $L(k)$ may be nonconvex, $X^o(k)$ is a convex polyhedral set. Further, by suitably defining $f(.,.)$ and k, we can have $\bar{x}$ lying in the interior of $X^o(k)$ with $X^o(k)$ containing no point in $X \cap S$. That is, any $z \epsilon X \cap S$ must satisfy $f(x,z) \geq k$ for at least one vector x of X. Since $f(x,z)$ is linear in z for a fixed X, we may use the basic disjunctive cut principle on this statement. Let us now illustrate the application of this to 0-1 integer programming. In this context, S is the set of integer valued points and currently, $\bar{x} \notin S$. Also, among other constraints, X contains the constraints $x_j \leq 1$ for each variable x_j. Given the simplex tableau representing the current point $\bar{x}$, we may identify the half lines

$$\xi^j = \{x: x = \bar{x} - a^j\lambda_j, \; \lambda_j \geq 0\}, \; j\varepsilon J \tag{6.30}$$

where J is the index set associate with n nonbasic variables. The function $f(.,.)$ which we select is given by

$$f(x,z): (x - \tfrac{1}{2} e)^t G(z - \tfrac{1}{2} e), \; (x,z) \; \varepsilon \; R^n \times R^n \tag{6.31}$$

where $e=(1,\ldots,1)$ and where

$$G = \begin{bmatrix} g_1 & & & 0 \\ & g_2 & & \\ & & \ddots & \\ 0 & & & g_n \end{bmatrix}, \; \sum_{i=1}^{n} g_i = n, \; g_i \geq 0, \; i=1,\ldots,n \tag{6.32}$$

and $g_i > 0$ for at least one $i\varepsilon\{1,\ldots,n\}$ for which $0 < \bar{x}_i < 1$. Further, we take the value of k to be $\frac{n}{4}$. Then, the set C is the level set

$$C = L(\tfrac{n}{4}) = \{x: (x - \tfrac{1}{2} e)^t G(x - \tfrac{1}{2} e) \leq \tfrac{n}{4}\} \tag{6.33}$$

This may be simplified to

$$C = L(\tfrac{n}{4}) = \{x: \sum_{i=1}^{n} g_i x_i (x_i - 1) < 0\} \tag{6.34}$$

Hence, all 0-1 points are contained in the boundary of the set $L(\frac{n}{4})$ while from (6.32), we observe that $\bar{x} \; \varepsilon \; \text{int } L(\frac{n}{4})$. Thus, int $L(\frac{n}{4})$ contains $\bar{x}$ but no point in $X \cap S$. Continuing, we have from Equation (6.29),

$$X^0(\tfrac{n}{4}) = \{z: (x - \tfrac{1}{2} e)^t G(z - \tfrac{1}{2} e) \leq \tfrac{n}{4} \text{ for each } x\varepsilon X\}$$

or

$$X^o\left(\frac{n}{4}\right) = \{z: \ (x - \frac{1}{2} \ e)^t Gz \leq \frac{1}{2} \ x^t Ge \quad \text{for each } x \epsilon X\} \tag{6.35}$$

One may show that int $X^o\left(\frac{n}{4}\right)$ contains $\bar{x}$ but no point in $X \cap S$. Hence, any point $z \ \epsilon \ X \cap S$ must satisfy at least one of the inequalities

$$(x - \frac{1}{2} \ e)^t G \ z \ \geq \frac{1}{2} \ x^t G \ e \ , \ \text{for } x\epsilon X. \tag{6.36}$$

Now, since $z \ \epsilon \ X$, from (6.30), we may write it as

$$z = \bar{x} - \sum_{j \epsilon J} a^j t_j \tag{6.37}$$

where t_j, $j \ \epsilon \ J$ are the current nonbasic variables. Substituting (6.37) into (6.36), we must have at least one of the following inequalities holding

$$\sum_{j \epsilon J} (\frac{1}{2} \ e \ - \ x)^t G \ a^j t_j \ \geq \frac{1}{2}(x + \bar{x})^t G \ e - x^t G \bar{x} \ , \ \text{for } x\epsilon X \tag{6.38}$$

But note that

$$\frac{1}{2}(x + \bar{x})^t G \ e - \ x^t G \bar{x} = \frac{1}{2} \ \Sigma (x_i + \bar{x}_i) g_i \ - \ \Sigma x_i \bar{x}_i g_i$$

$$> \frac{1}{2} \ \Sigma (x_i^2 + \bar{x}_i^2) g_i \ - \ \Sigma x_i \bar{x}_i g_i \ = \frac{1}{2} \ \Sigma (x_i - \bar{x}_i)^2 g_i \geq 0 \tag{6.39}$$

since $x_i^2 \leq x_i$, $\bar{x}_i^2 \leq \bar{x}_i$, $g_i \geq 0$ for each $i=1,\ldots,n$ and $g_i > 0$ for some i for which $0 < \bar{x}_i < 1$ (see (6.32)).

Thus, the right hand side of (6.38) is positive and we may normalize (6.38) by its right hand side for each $x\epsilon X$. Thus, applying Theorem 3.1, Equation (3.16), we may derive the disjunctive cut

125

$$\sum_{j \in J} (1/\bar{\lambda}_j) t_j \geq 1$$

where, $\bar{\lambda}_j$, $j \in J$ is given by

$$(1/\bar{\lambda}_j) = \left[\max_{x \in X} \frac{(\frac{1}{2} e - x)^t G a^j}{\frac{1}{2} (x+\bar{x})^t Ge - x^t G\bar{x}} \right], \quad j \in J \qquad (6.40)$$

Now, let us determine the intersection cut based on $X^o(\frac{n}{4})$ of (6.35). This cut is given as

$$\sum_{j \in J} (1/\hat{\lambda}_j) t_j \geq 1 \qquad (6.41)$$

where

$$\hat{\lambda}_j = \max\{\lambda_j : z = (\bar{x} - a^j \lambda_j) \in X^o(\frac{n}{4})\}$$

$$= \max\{\lambda_j : (x - \frac{1}{2} e)^t G(\bar{x} - a^j \lambda_j) \leq \frac{1}{2} x^t Ge \text{ for each } x \in X\}$$

$$= \max\{\lambda_j : \lambda_j (\frac{1}{2} e - x)^t Ga^j \leq \frac{1}{2}(x + \bar{x})^t Ge - x^t G\bar{x} \text{ for each } x \in X\}$$

But noting (6.39), we obtain

$$\hat{\lambda}_j = \max\{\lambda_j : \lambda_j \left[\frac{(\frac{1}{2} e - x)^t Ga^j}{\frac{1}{2}(x + \bar{x})^t Ge - x^t G\bar{x}} \right] \leq 1 \text{ for each } x \in X\}$$

or

$$\hat{\lambda}_j = \max\{\lambda_j : \lambda_j \max_{x \in X} \left[\frac{(\frac{1}{2} e - x)^t Ga^j}{\frac{1}{2} (x + \bar{x})^t Ge - x^t G\bar{x}} \right] \leq 1\} \qquad (6.42)$$

Thus, if $(1/\bar{\lambda}_j) > 0$ in (6.40), then from (6.42), we observe that $1/\hat{\lambda}_j = 1/\bar{\lambda}_j$. On the other hand, if $1/\bar{\lambda}_j \leq 0$ in (6.40), then $\hat{\lambda}_j \to \infty$ or $1/\hat{\lambda}_j = 0$. Hence, the cut $\sum_{j \in J} \dfrac{1}{\bar{\lambda}_j} t_j \geq 1$ uniformly dominates $\sum_{j \in J} \left(\dfrac{1}{\hat{\lambda}_j}\right) t_j \geq 1$ and in fact, the former (disjunctive) cut implies the latter (intersection) cut.

Before concluding, we note that polar cut results analogous to those given for 0-1 programming above, may be obtained for (nonconvex) quadratic programming problems as well.

In the final two chapters of these notes, we will consider some special cases of disjunctive programs. To begin with, in the next chapter, we will treat facial disjunctive programs as introduced in Chapter V. Specifically, we will discuss two finitely convergent schemes for solving such problems. Thereafter, in Chapter VIII, we will dwell briefly on specific applications of some of the classes of disjunctive programs introduced in Chapter I.

6.5 Notes and References

In view of Theorem 2.1, any valid cut for a disjunctive program should be recoverable or can be dominated by a disjunctive cut of the theorem. Balas [4,6] has discussed this relationship in some detail for integer and nonlinear programs. As noted by Balas [4,6] and Glover [18,19], the new cuts clearly have the capability of improving some of the well-known cuts, and this is demonstrated by the discussion in the chapter.

Chapter VII

FINITELY CONVERGENT ALGORITHMS FOR FACIAL DISJUNCTIVE PROGRAMS WITH APPLICATIONS TO THE LINEAR COMPLEMENTARITY PROBLEM

7.1 Introduction

In our discussion of Chapter V, we had introduced a special class of disjunctive programs called facial disjunctive programs, examples of which included the zero-one linear integer programming problem and the linear complementarity problem. We had seen that for this special class of problems, it was relatively easy to generate the convex hull of feasible points. In this chapter, we will discuss two finitely convergent schemes which solve facial disjunctive programs by generating facets of the convex hull of feasible points as and when needed, until such time as either a suitable termination criterion is met or the problem is solved through the generation of the entire convex hull.

The organization of this chapter is as follows. We first briefly discuss how Theorem 5.11 may be exploited to develop a finite scheme for facial disjunctive programs. Thereafter, we present in greater detail, a second alternative method based on Theorem 5.10. This technique is the principal thrust of this chapter. Finally, we demonstrate how this latter method may be specialized for the linear complementarity problem.

7.2 Principal Aspects of Facial Disjunctive Programs

For the sake of completeness and convenience, let us re-introduce certain notations and concepts to be used in this chapter. The facial disjunctive program under consideration is

<u>FDP</u>:   minimize $c^t x$

subject to $x \varepsilon X = \{x: Dx = d, x \geq 0\}$ (7.1)

$$x \varepsilon Y = \bigcap_{h \varepsilon H} [\bigcup_{i \varepsilon Q_h} \{x: a_i^h x \geq b_i^h\}]$$ (7.2)

Here, c is a lxn real vector, $x=(x_1,\ldots,x_n)$ is a (nx1) vector of variables, X is assumed to be a non-empty and bounded polyhedral set (regularized as assumed in Chapter V, if necessary). Further, D is an mxn real matrix and d is an nx1 real vector. The set Y is a conjunction of $|H| < \infty$ disjunctions, with $H=\{1,\ldots,h\}$, say. Observe that we have deviated in consistency as regards to notation in this chapter so as to make the presentation more readable. Since this chapter is written to be basically self contained, we hope that this will not lead to any confusion. Continuing, the set Y defines for each $h \varepsilon H$, a disjunction which states that at least one of the constraints $a_i^h x \geq b_i^h$ must be satisfied for some $i \varepsilon Q_h$. Here, a_i^h is a lxn real vector and b_i^h is a real scalar for each $i \varepsilon Q_h$, $h \varepsilon H$. The constraint index sets Q_h, $h \varepsilon H$ may contain common elements correponding to common constraints, and are otherwise disjoint.

Recall from Chapter V that the disjunction $x \varepsilon \bigcup_{i \varepsilon Q_h} \{a_i^h x \geq b_i^h\}$ is called <u>facial</u> with respect to X if $X \cap \{x: a_i^h x \geq b_i^h\}$ is a face of X for each $i \varepsilon Q_h$. In addition, the disjunctive program FDP is said to be <u>facial</u> if each of the disjunctions $h \varepsilon H$ is facial with respect to X. As before, by a <u>face</u> of X we imply a subset of X defined by the intersection of X with a hyperplane which supports it.

Now, with our assumption of X being a bounded polyhedral set and with Y as specified in (7.2), we have,

$$F = c\ell \text{ conv } X \cap Y = \text{conv } X \cap Y$$ (7.3)

Further, let us inductively define

$$K_0 = X$$

$$K_h = \text{conv}[\bigcup_{i \varepsilon Q_h} (K_{h-1} \cap \{x: a_i^h x \geq b_i^h\})] \text{ for } h=1,\ldots,\hat{h}$$ (7.4)

Then, Theorem 5.11 and 5.10 are respectively re-stated below as properties P1 and P2.

P1: $K_{\hat{h}}$ of Equation (7.4) is equal to F of Equation (7.3)

P2: {Extreme points of F of Equation (7.3)} $\subseteq$ {Extreme points of X of Equation (7.1)}

We will now proceed to discuss the skeleton of a procedure for solving Problem FDP based on property P1 above. This discussion will also serve to lay the foundations for the second procedure which is treated at length in this chapter.

7.3 Stepwise Approximation of the Convex Hull of Feasible Points

Essentially, this scheme for solving Problem FDP is a relaxation strategy. To begin with, the constraints (7.2) are relaxed and the resulting linear program is solved. If the optimal solution $\bar{x}$, say, satisfies $\bar{x} \in Y$, then this solution is also optimal to FDP. Otherwise, a disjunction for some $h \in H$ is violated. Based on a violated disjunction, a cutting plane which deletes $\bar{x}$ but no point of X satisfying this disjunction and hence, no point of F, is now generated. This cutting plane is imposed as an additional constraint and the optional solution $\bar{x}$ is hence updated. This process is repeated till can optimal solution to some relaxed problem is feasible to (7.2).

Finiteness of the scheme is based on a result which is basically a strengthened version of the reverse part of Theorem 2.1, namely, the fundamental disjunctive cut principle. The result is stated below (without proof).

Theorem 7.1

Let $S_r = \{x: A^r x \geq b^r, x \geq 0\}$ for each $r \in R$ be non-empty sets and consider the disjunctive $x \in \bigcup_{r \in R} S_r$. Further, let $|R| = \tau$, say, and define the set

$$
\begin{aligned}
E = \{(\lambda^1, \ldots, \lambda^\tau, \alpha, \alpha_0): \ & \lambda^r A^r - \alpha = 0 && \text{for } r=1,\ldots,\tau \\
& \lambda^r b^r - \alpha_0 \geq 0 && \text{for } r=1,\ldots,\tau \\
& \sum_r \sum_i \lambda_i^r = 1 \\
& \lambda^r \geq 0 \ \text{ for } r=1,\ldots,\tau \}
\end{aligned}
$$

$$(7.5)$$

where for each $r=1,\ldots,\tau$, the vector λ^r has as many columns as A^r has rows, and where α is of the same dimension as x, with α_0 being a scalar. Then,

$$c\ell \ \text{conv} \ \bigcup_{r\in R} S_r = \{x: \ \alpha^i x \geq \alpha_{i0} \ \text{for each i such that}$$

$$(\lambda^{1i},\ldots,\lambda^{\tau i},\alpha^i,\alpha_{i0}) \ \text{is an extreme point of E}\} \qquad (7.6)$$

In other words, if we had an enumeration of the extreme points of E of the form $(\lambda^{1i},\ldots,\lambda^{\tau i},\alpha^i,\alpha_{i0})$ indexed by i, then we could construct the closure of the convex hull of points feasible to the disjunction $x \in \bigcup_{r\in R} S_r$ as the intersection of the half-spaces $\alpha^i x \geq \alpha_{i0}$. This fact along with Property P1 may be used roughly as follows.

Initially, let us solve the problem of minimizing cx over the set $K_0 \equiv X$. Assume for the sake of simplicity that the optional solution $\bar{x}$ found violates the disjunction corresponding to $h=1$. Then, $\bar{x} \notin K_1$ and one may derive a cut corresponding to an extreme point of E of Equation (7.5) which deletes $\bar{x}$. Here, the constraint sets $A^r x \geq b^r$, for $r \in R$ correspond to $K_0 \cap \{x: a_i^h x \geq b_i^h\}$ for $i \in Q_1$. The cut may be simply derived by maximizing $\alpha_0 - \alpha x$ over this set E. Now, during the course of the procedure, whenever the disjunction for $h=1$ is violated, this step may be repeated. Clearly, from Theorem 7.1, this can happen only finitely often, the entire set K_1 being constructed in the worst case. In a similar manner, one may inductively argue that subsequent disjunction violations considered can be repeated only a finite number of times. Again, assuming for the sake of simplicity that these disjunction violations occur and are considered in the order $h=1,2,\ldots$ one may note that when deriving cuts for the jth disjunction, the constraints $A^r x \geq b^r$ for $r \in R$ used in the set E of Equation (7.5) correspond to the intersection of the set K_0, the cuts generated for the disjunction violations $1,2,\ldots,j-1$ and the disjunctive constraints indexed by Q_j. For algorithmic purposes, whenever an updated solution violates more than one disjunction which has been previously considered, the cut derived is based on the most recent one of these disjunctions. In this manner, at worse, one would construct the sets $K_0, \ldots, K_{\hat{h}}$ in their entirety. Typically, the actual sequence of sets constructed may be only an approximation of these sets in the vicinity of an optional solution.

7.4 Approximation of the Convex Hull of Feasible Points through an Extreme Point Characterization

The second procedure (which we shall call the Extreme Point Method) for solving Problem FDP is basically the same type of relaxation scheme as discussed in the foregoing section. Hence, a series of cutting planes and updated solutions to the relaxed problems are generated till such time as an updated solution is found which satisfies the disjunction (7.2). Whereas we had to specify restrictions on the type of and manner in which the cutting planes were generated in the previous section in order to ensure finiteness, we have some flexibility in this respect in the present approach. Instead, in order to invoke Property P2, we place specific restrictions on the type of points at which the cuts are generated. Specifically, these points are required to be so called extreme faces of the set X with respect to cuts generated at any stage of the procedure. This concept of extreme faces is discussed in the following subsection.

7.4.1 Extreme Faces and Their Detection

Let us assume that at a particular stage s cuts, $Gx \leq g$, have been generated in the space of the x-variables. Let

$$\Lambda = \{x \in R^n : Gx + Ix_s = g, \; x_s \geq 0\} \qquad (7.7)$$

be the subset of R^n feasible to these cuts. Here, $x_s = (x_{n+1}, \ldots, x_{n+s})$ denotes the vector of slack variables (with the superscript t being used to designate the matrix transpose operation), and I is an identity matrix of size s. Further, let $N = \{1, \ldots, n\}$ denote the index set of the original x-variables, which we will call key variables. Also, let $S = \{n+1, \ldots, n+s\}$ denote the index set of the slack variables for the s cuts, which we will call as nonkey variables. For a set $Z \subseteq N$, let

$$F_Z = \{x \varepsilon X: x_j = 0 \text{ for } j \varepsilon Z\} \qquad (7.8)$$

Note that all faces of X can be represented as F_Z for some suitable set Z. Finally, for any point $x \varepsilon F_Z$, let the zero components of x be denoted by

$$Z(x) = \{j \varepsilon N: x_j = 0\} \qquad (7.9)$$

<u>Definition 4.1</u>

Let F_Z be a face of X defined by some $Z \subseteq N$ such that $F_Z \cap \Lambda \neq \phi$. Then F_Z is an <u>extreme face</u> of X relative to Λ if for any two points $x^1, x^2 \varepsilon F_Z \cap \Lambda$, we have $Z(x^1) = Z(x^2)$.

In other words, an extreme face F_Z satisfies the property that $F_Z \cap \Lambda$ does not contain any point in a lower dimensional face of X. Examples of extreme faces of X relative to Λ are extreme points of X feasible to Λ, or an edge of X not disjoint with Λ but with neither of the two extreme points of X defining this edge being feasible to Λ.

Now, observe that Property P2 essentially directs that the search for an optimal solution to Problem FDP may be restricted to a search among the extreme points of X. However, we will find it simpler to restrict our search to a larger set, namely, the extreme faces of X. Since extreme faces of X relative to some Λ are also faces of X, the number of such extreme faces of X (relative to all Λ's) is finite. Hence, a procedure which detects and deletes in a finite number of steps at least one extreme face per iteration is finitely covergent. This is indeed the principal thrust of the present scheme.

Given a simplex tableau representation of an extreme point of $X \cap \Lambda$ at any stage, a simple procedure to find an extreme face of X relative to Λ utilizes the following <u>restricted basis entry rule</u>:

"Only a nonkey variable x_j, $j \varepsilon S$, is eligible to enter the basis" $\qquad (7.10)$

Based on this, the method outlined below either finds an extreme face or indicates that no such face exists.

Step 1

Let x_r denote the largest valued basic key variable in the current solution which has not yet been considered at a previous iteration. If no such variable exists, go to Step 3. Otherwise, proceed to Step 2.

Step 2

Solve the Problem P_r: minimize $\{x_r: x \in X \cap \Lambda\}$ as a linear program subject to the restricted basis entry rule (7.10). If the solution yields $x_r = 0$ and x_r is basic, pivot it out of the basis, if possible, by exchanging it with a nonkey, nonbasic variable. Return to Step 1.

Step 3

If all key variables are basic, there is no extreme face of X relative to Λ. Otherwise, the current set Z of indices of nonbasic key variables defines, through Equation (7.8), an extreme face F_Z of X relative to Λ. In particular, if all nonbasic variables are key variables, then F_Z represents an extreme point of X.

7.4.2 Schema of the Proposed Approach

The procedure we adopt operates as follows. At any stage, given the set Λ of Equation (7.7), we solve the relaxes problem

$$P(\Lambda): \quad \text{minimize} \quad \{cx: x \in X \cap \Lambda\} \quad\quad (7.11)$$

If an optimal solution $\bar{x}$ to this problem satisfies $\bar{x} \in Y$ of Equation (7.2), we stop wtih $\bar{x}$ as an optimal solution to Problem FDP. Otherwise, we generate a cut based on a violated disjunction, and then after updating the tableau, we use the routine of Section 7.4.1 to find an extreme face F_Z of X relative to Λ. If no extreme faces exist, then we terminate with the current best known solution as optimal to FDP. Otherwise, depending on the dimension of F_Z, two possible routes are open to us. If F_Z is of dimension greater than zero, then a disjunctive face cut is developed which deletes F_Z but no extreme point of X feasible to Λ. Details of such a cut are presented in Section 7.5. On the other hand, if F_Z represents an extreme point of X then we check if this extreme point is feasible to Y. If it is, then we update the current best known solution, if necessary, and again generate a disjunctive face cut which deletes only this particular extreme point of X. If the extreme point is infeasible to Y, however, a stronger disjunctive cut may be developed as discussed in Chapter IV and re-iterated in Section 7.5. In any case, after the appropriate cut has been generated and Λ has been updated, we say that an iteration has been completed. A new iteration is now commenced by solving Problem $P(\Lambda)$ of Equation (7.11).

As an additional expedient, we will also impose the cost cut

$$c^t x \leq \hat{v}$$

where $\hat{v}$ is the current best known objective value of Problem FDP. Hence, the right hand side of this cut is simply updated each time an improved solution is detected. Although this cut will not affect the solution of Problems $P(\Lambda)$, it will assist in confining the search to improving solutions during the extreme face finding routine. This is essential because otherwise, the extreme face finding routine would simply concentrate on feasibility, regardless of objective function values.

Figure 7.1 gives a flow chart of the proposed scheme. The collection of extreme faces of X relative to all possible sets Λ being finite, this method is clearly finitely convergent. Instead of reviewing in detail the general concepts involved in generating disjunctive face cuts, we discuss its generation for the linear complementarity problem in particular and merely allude subsequently to the extension of this to the general case.

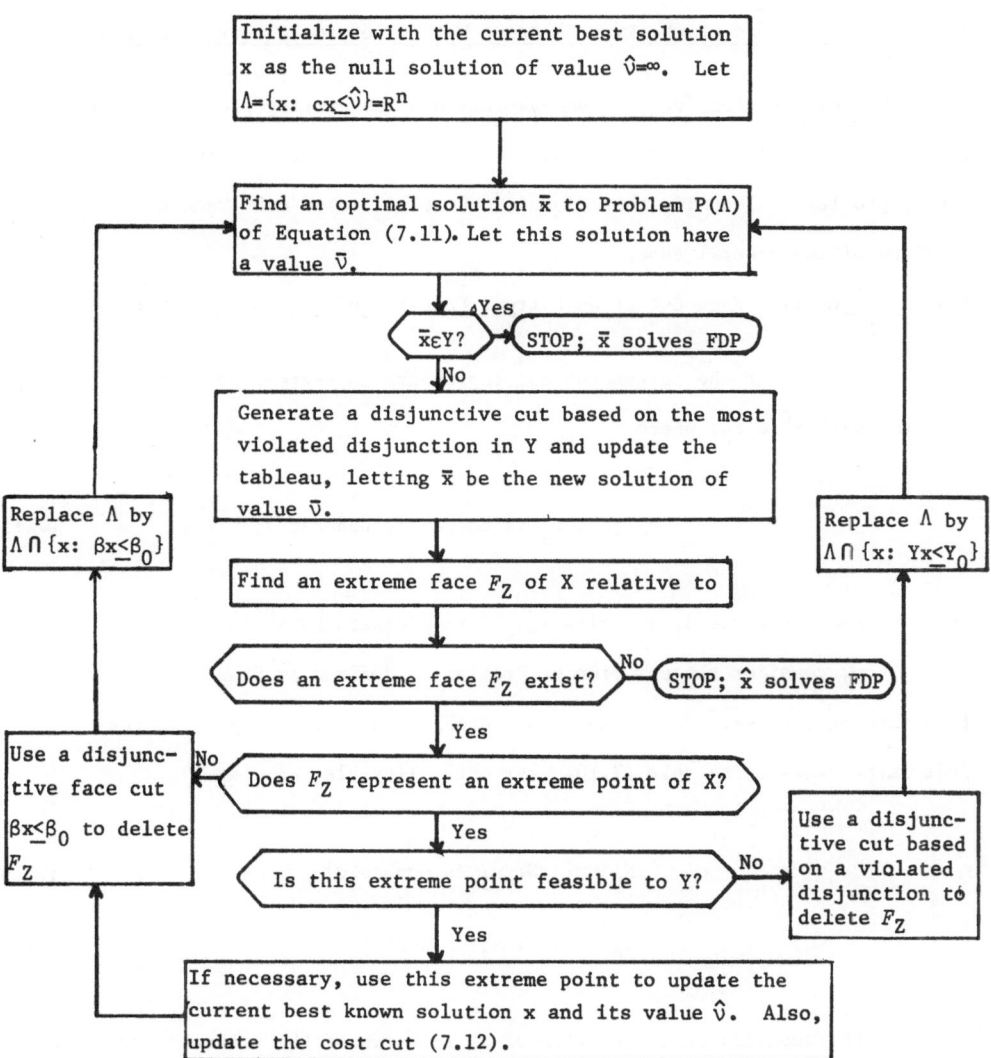

Figure 7.1. Flow-Chart for the Proposed Scheme

7.5 Specializations of the Extreme Point Method for the Linear Complementarity Problem

In this section, we will demonstrate how the cutting planes to be used in the procedure depicted in Figure 7.1 may be generated for the linear complementarity problem. Alongside this discussion, we will also make remarks for the handling of the general case.

7.5.1 Disjunctive Face Cut at an Extreme Face F_Z Which is not an Extreme Point of X

Suppose that the current tableau represents an extreme point $x^0 = (x_1^0, \ldots, x_n^0)$ of $X \cap \Lambda$ with $x^0 \in F_Z$, where

$$Z = \{j \in N: x_j \text{ is currently nonbasic}\} \tag{7.13}$$

Let us assume that the disjunction $x_p x_q = 0$ is violated by x^0.

Now, consider the solution of Problem P_p defined in Step 2 of the extreme face finding routine. Recall that this problem is solved subject to the restricted basis entry rule (7.10). At optimality, let

$$N_p = \{j \in N: x_j \text{ is nonbasic}\} \tag{7.14}$$

$$S_p = \{j \in S: x_j \text{ is nonbasic}\} \tag{7.15}$$

and let the canonical representation of x_p in terms of the nonbasic variables x_j, $j \in N_p \cup S_p$ be

$$x_p + \sum_{j \in N_p} a_{pj} x_j + \sum_{j \in S_p} a_{pj} x_j = b_p$$

Since $N_p \subseteq Z$, by adding suitable zero coefficients, the above equation may be written as

$$x_p + \sum_{j \in Z} a_{pj} x_j + \sum_{j \in S_p} a_{pj} x_j = b_p \tag{7.16}$$

In a similar manner, after solving P_q, we would have an equation

$$x_q + \sum_{j \varepsilon Z} a_{qj} x_j + \sum_{j \varepsilon S_q} a_{qj} x_j = b_q \tag{7.17}$$

It is easy to show that

$$a_{pj} \leq 0 \quad j \varepsilon S_p, \; a_{qj} \leq 0 \quad j \varepsilon S_q, \; b_p > 0, \; b_q > 0 \tag{7.18}$$

Now, the requirement that at least one of $x_p \leq 0$, $x_q \leq 0$ must hold may be written as requiring that at least one of the following constraint sets must be satisfied

$$\sum_{j \varepsilon Z \cup S_{pq}} \left(\frac{a_{rj}}{b_r} \right) x_j \geq 1; \; x_j \geq 0 \text{ for } j \varepsilon Z \cup S_{pq} \quad \text{for } r=p,q \tag{7.19}$$

where $S_{pq} = S_p \cup S_q$ and again, we have suitably defined zero coefficients wherever necessary. From Theorem 2.1, a valid cut is

$$\sum_{j \varepsilon Z \cup S_{pq}} \left(\max\left\{ \frac{a_{pj}}{b_p}, \frac{a_{qj}}{b_q} \right\} \right) x_j \geq 1 \tag{7.20}$$

Observe from (7.18) that (7.20) implies

$$\sum_{j \varepsilon Z} \left(\max\left\{ \frac{a_{pj}}{b_p}, \frac{a_{qj}}{b_q} \right\} \right) x_j \geq 1$$

and hence, (7.20) deletes F_Z since any $x \varepsilon F_Z$ satisfies $x_j = 0$ for $j \varepsilon Z$. Finally, note that either in the general case of facial disjunctive programs or in the present application when $x^0 \varepsilon Y$, one may obtain an equation of the type (7.16) for each $r \varepsilon N$ such that $x_r^0 > 0$ and then derive a cut based on the disjunction that at least one of these variables x_r must be zero at any extreme point of x feasible to Λ, if such a point exists.

7.5.2 Disjunctive Cut at an Extreme Face F_Z which is an Extreme Point of X

Let x^0 be the extreme point of X represented by the current tableau as the extreme face F_Z. Again, if $x^0 \in Y$, then we develop a disjunctive face cut as in the general case of the foregoing discussion. On the other hand, if some disjunction $x_p x_q = 0$ is violated, then a deeper cut may be generated in the following manner.

Let the canoical representation of the (positive) basic variables x_p and x_q in the current tableau be given by

$$x_p + \sum_{j \epsilon Z} a_{pj} x_j = b_p > 0$$

$$x_q + \sum_{j \epsilon Z} a_{qj} x_j = b_q > 0$$

(7.21)

where Z is currently also the index set for nonbasic variables. Now, the disjunction that at least one of the variables x_p, x_q equals zero may be written as the requirement that at least one of the constraint sets

$$\left\{ \sum_{j \epsilon Z} \left(\frac{a_{pj}}{b_p} \right) x_j \geq 1, \ x_j \geq 0 \text{ for } j \epsilon Z \right\}, \left\{ \sum_{j \epsilon Z} \left(\frac{a_{qj}}{b_q} \right) x_j \geq 1, \right.$$

$$\left. x_j \geq 0 \text{ for } j \epsilon Z \right\}$$

(7.22)

must be satisfied. Through Theorem 2.1, a valid cut based on this statement is

$$\sum_{j \epsilon Z} \bar{\pi}_j x_j \geq 1$$

(7.23)

where,

$$\bar{\pi}_j = \max\left\{ \frac{a_{pj}}{b_p}, \frac{a_{qj}}{b_q} \right\} \text{ for each } j \epsilon Z$$

(7.24)

which clearly deletes $F_Z = x^0$. Now, in Chapter IV, we discussed how this cut may further strengthened by considering nonnegativity conditions on the other basic variables also. Hence, if we let B_{pq} be the index set of basic variables x_r, $r \neq p$, $r \neq q$ then we may replace the disjunction (7.22) by the following disjunction, where we have used a canonical representation for the basic variables x_r, $r \epsilon \beta_{pq}$ similar to that in (7.21). This disjunction states that at least one of the following constraint sets must be satisfied

$$\{ \sum_{j \epsilon Z} a_{rj}x_j \leq b_r \text{ for } r \epsilon B_{pq}, \sum_{j \epsilon Z} a_{pj}x_j \geq b_p, x_j \geq 0 \text{ for } j \epsilon Z\}$$

$$\tag{7.25}$$

$$\{ \sum_{j \epsilon Z} a_{rj}x_j \leq b_r \text{ for } r \epsilon B_{pq}, \sum_{j \epsilon Z} a_{pj}x_j \geq b_q, x_j \geq 0 \text{ for } j \epsilon Z\}$$

The improvement technique proposed in Chapter IV essentially attempts to derive a cut in terms of the nonbasic variables x_j, $j \epsilon Z$ such that this cut is a support for the closure of the convex hull of the union of the two sets in (7.25). The method accomplishes this by commencing with the cut (7.23), say, and attempting to improve (decrease) as much as possible each cut coefficient one at a time, holding the other cut coefficients fixed. Hence, if at any stage, if (7.23) represents the current cut and if one is trying to reduce the coefficient of x_k, then as in Chapter IV this coefficient is given by the larger of the optimal values of the two linear programs LP_{kp} and LP_{kq}, where

$$LP_{kh}: \quad \text{maximize} \quad \xi - \sum_{\substack{j \epsilon Z \\ j \neq k}} \bar{\pi}_j y_j$$

$$\text{subject to} \quad \sum_{\substack{j \epsilon Z \\ j \neq k}} a_{rj}y_j - b_r\xi \leq -a_{rk} \quad \text{for } r \epsilon B_{pq}$$

$$b_h\xi - \sum_{\substack{j \epsilon Z \\ j \neq k}} a_{hj}y_j \leq a_{hk} \tag{7.26}$$

$$\xi \geq 0, y_j \geq 0 \text{ for } j \epsilon Z - \{k\}$$

Again, as indicated in Chapter IV, both LP_{kp} and LP_{kq} need not necessarily be solved independently. Thus, for instance, if Equation (7.24) yields $\bar{\pi}_k = \frac{a_{pk}}{b_p}$, say, then one may solve LP_{kp} first. If the optimal value $\bar{\pi}_{kp}$ of LP_{kp} equals $\bar{\pi}_k$, then LP_{kq} need not be solved. Otherwise, LP_{kq} may be solved with the added constraint that its objective value exceeds $\bar{\pi}_{kp}$.

Of course, to reduce the effort in the generation of such cuts, one need not include all the constraints for $r \epsilon B_{pq}$ in the sets of Equation (7.25). Instead, some heuristic rule made be used to select a subset of these contraints. For example, one may select those constraints from B_{pq} which delete at least one of the finite intercepts which the cut (7.23) makes on some axis. Hence, one may select

$$\{ r\epsilon B_{pq} : \frac{a_{rk}}{\bar{\pi}_k} > b_r \text{ for some } k \epsilon \bar{\pi}_k \neq 0 \}$$

7.6 Notes and References

The term "facial disjunctive programs" was first used by Balas [5] where the principal result, namely Properties P1 and P2, discussed in this chapter, are proved. Those results have led to two finitely convergent algorithms for facial programs presented in this chapter. The first is due to Jeroslow [26] and makes use of one of the properties. The second algorithm is based on the other property and uses the concept of "extreme faces" first presented by Majthay and Whinston [28].

CHAPTER VIII

SOME SPECIFIC APPLICATIONS OF DISJUNCTIVE PROGRAMMING PROBLEMS

8.1 Introduction

In Chapter I we discussed in general the major applications of disjunctive programming problems. These included the generalized lattice point and related problems, the cardinality constrained problem, the extreme point programming problem and the binary mixed integer linear programming problem. In this chapter, we will present some specific applications which are subsumed under these general classes of problems.

8.2 Examples of Bi-Quasiconcave Problems

The Bi-Quasiconcave problem may be written as

$$\text{minimize } \{f(x,y): x \epsilon S_x, y \epsilon S_y\}$$

where S_x and S_y are polyhedral sets in variables x and y respectively and (.,.) is a real valued function such that $f(.,y)$ and $f(x,.)$ are quasiconcave for any fixed x and y. It is easy to see that this latter property guarantees that an optimal solution is obtained at an extreme point of $S_x \times S_y$. Thus these problems are essentially extreme point optimization problems. We will now discuss some practical Bi-Quasiconcave problems.

8.2.1 Orthogonal Production Scheduling – A Multiperiod Activity Analysis Model

Let us first of all consider the class of problems known as the Multiperiod Activity Analysis problems or the multistage production problems. These problems have the mathematical form

$$\text{minimize} \quad \sum_{k=1}^{K} (c^k)^t x^k$$

$$\text{subject to} \quad Ax^k \geq b^k \text{ for } k=1,\ldots,K$$

$$x^k \geq 0$$

Here, we have, say n activities producing m commodities over K periods. Thus, x^k is a vector representing the activity levels at period k, k=1,...,K, for activities 1,...,n, say. Further, A_{mxn} is the matrix of input-output or technological coefficients, b^k is a vector which denotes the requirements for various commodities 1,...,m in period k and c^k is a vector which represents the unit cost associated with each activity 1,...,n in period k.

However, certain physical considerations may require that certain orthogonal constraints of the form $x_j^{k-1} \cdot x_j^k = 0$, k=2,...,K hold for some activities j. For example, this may arise in the context of machine scheduling wherein due to maintenance considerations, certain activities cannot be scheduled in two consecutive periods. As another example, they may arise in an agricultural production situation wherein certain crops cannot be raised in two consecutive periods to preserve specific nutrients in the soil.

This problem may be transformed into a Bi-Quasiconcave Programming problem as follows. Let us assume, merely for convenience, that each activity is restricted by the orthogonal scheduling constraint mentioned above. Then we may use the penalty function method to ascribe a high cost to any schedule which is infeasible to these orthogonal constraints. That is, letting M be a large constant, we may formulate the orthogonal production problem as

$$\text{minimize} \quad \sum_{k=1}^{K} (c^k)^t x^k + M \sum_{k=2}^{K} (x^{k-1})^t x^k$$

$$\text{subject to} \quad Ax^k \geq b^k, \quad k=1,...,K$$

$$x^k \geq 0$$

For the case K=2, this problem is clearly a Bi-Quasiconcave Program; in fact, for K=2, it is a Bilinear Programming Problem with an optimal solution being an extreme point of $X_1 \times X_2$ where $X_1 = \{x^1 \geq 0: Ax^1 \geq b^1\}$, $X_2 = \{x^2 \geq 0: Ax^2 \geq b^2\}$.

143

8.2.2 Application to Game Theory

Consider a two player game where player P_1 selects his strategy first as a vector x from the set $X = \{x: A_1x \leq b, x \geq 0\}$. Depending on the strategy x selected by P_1, let us say that player P_2 selects a strategy y from the set $Y(x) = \{y: A_2y \leq d + Cx, y \geq 0\}$. Here, Y(x) is assumed bounded and nonempty for each $x\epsilon X$. Further, let us say that when P_1 selects strategy x and P_2 selects strategy y, there is an associated payoff $f(x,y) = p^tx + q^ty$ from P_1 to P_2, where p and q are given cost vectors. Thus, given x, P_2 will solve the problem

$$\text{maximize} \quad q^ty$$
$$\text{subject to} \quad y\epsilon Y(x)$$

Let y(x) denote an optimal solution to the above problem. Hence, knowing the technique to be adopted by P_2, P_1 will try to select a strategy $x\epsilon X$ which minimizes f(x,y(x)), that is, he will solve the problem

$$\text{minimize} \quad \{p^tx + \text{minimum}\{(d + Cx)^tz: A_2z \geq q, z \geq 0\}\}$$
$$\text{subject to} \quad A_1x \leq b, x \geq 0$$

where $q^ty(x) = \max\{q^ty: A_2y \leq d + Cx, y \geq 0\}$ has been rewritten as $q^ty(x) = \min\{(d + Cx^tz: A_2z \geq q, z \geq 0\}$. Hence, the above problem may equivalently be written as

$$\text{minimize} \quad d^tz + p^tx + z^tCx$$
$$\text{subject to} \quad A_2z \geq q, z \geq 0$$
$$A_1z \leq b, x \geq 0$$

This problem is again a Bi-Quasiconcave Programming Problem; in fact, it is a bilinear problem.

8.2.3 Multi-Stage Assignment Problem

For the sake of simplicity, consider a two-stage assignment problem. The development given below may easily be generalized to the multi-stage problem. Hence, suppose we have N jobs and N machines with the stipulation that at each of the two stages, one and only one machine should be assigned to each job. The profit of assigning machine i to job k at the first stage is simply p_{ik}. However, the profit of assigning job i to machine j at the second stage depends on the job k to which machine i was assigned at the first stage. This profit is accordingly given by $\bar{p}_{ij} + q_{ijk}$. Thus, the total two-stage profit is given by

$$\sum_{i=1}^{N} \sum_{j=1}^{N} p_{ij} x_{ij}^1 + \sum_{i=1}^{N} \sum_{j=1}^{N} \bar{p}_{ij} x_{ij}^2 + \sum_{i=1}^{N} \sum_{j=1}^{N} \sum_{k=1}^{N} q_{ijk} x_{ij}^2 x_{ik}^1$$

where

$$x^r \in X_r = \{x^r : \sum_{i=1}^{N} x_{ij}^r = 1, \; j=1,\dots,N, \; \sum_{j=1}^{N} x_{ij}^r = 1, \; i=1,\dots,N, \; x_{ij}^r = 0, 1$$

for i, j$\in\{1,\dots,N\}$ and for r=1,2

That is X_1 and X_2 represent the assignment constraints at stages one and two respectively. Hence defining N^2 vectors $p = (p_{ij})$, $\bar{p} = (\bar{p}_{ij})$ and letting Q be an appropriate matrix made up of zeroes and coefficients q_{ijk}, we may formulate this problem as

$$\begin{aligned} \text{minimize} \quad & p^t x^1 + \bar{p}^t x^2 + (x^1)^t Q x^2 \\ \text{subject to} \quad & x^1 \in X_1 \\ & x^2 \in X_2 \end{aligned}$$

This is again a bilinear programming problem.

8.2.4 Rectilinear Distance Location-Allocation Problems

As a final example of Bi-Quasiconcave Programming Problem, we consider this problem which is again a Bilinear Programming Problem. Specifically, consider a multifacility location-allocation problem which involves the distribution of several products among some new facilities to be located and between these new and other already existing facilities. Thus, suppose n new facilities are to be located. Let the variables denoting their location in a two-dimensional layout be (x_i, y_i), $i=1,\ldots,n$. Further, let there be m existing facilities currently located at (x_i, y_i), $i=n+1,\ldots,n+m$. Let a_{ik} denote the availability of product k, $k=1,\ldots,p$, say, at a new facility i, for $i=1,\ldots,n$ and let b_{ik} denote the requirement of product k at a new or existing facility i, $k=1,\ldots,p$, $i=1,\ldots,n+m$. Let us also assume that each unit of product k supplied from new facility i to new or existing facility j costs c_{ijk} with a corresponding transportation cost per unit distance of t_{ijk}. Here, the distances are taken to be measured using the rectilinear norm. This distance measure is appropriate in the context of movement along city streets or in a grid of aisles in a factory or a warehouse.

The problem is to determine the locations (x_i, y_i) for the new facilities $i=1,\ldots,n$ and to find feasible allocations u_{ijk} of product k from new facility i to new or existing facility j so as to minimize the total purchase (or manufacture) and transportation costs. Mathematically, this problem may be written as

$$\text{minimize} \quad \sum_{k=1}^{p} \sum_{i=1}^{n} \sum_{j=1}^{n+m} \{c_{ijk} + t_{ijk}(|x_i - x_j| + |y_i - y_j|)\} u_{ijk}$$

$$\text{subject to} \quad u \in U = \{u = (u_{111},\ldots,u_{n,n+m,p}):$$

$$\sum_{j=1}^{n+m} u_{ijk} \le a_{ik} \quad \text{for } i=1,\ldots,n, \ k=1,\ldots,p$$

$$\sum_{i=1}^{n} u_{ijk} = b_{jk} \quad \text{for } j=1,\ldots,n+m; \ k=1,\ldots,p$$

$$u_{ijk} \ge 0 \quad \text{for } i=1,\ldots,n; \ j=1,\ldots,n+m, \ k=1,\ldots,p\}.$$

One may now use the usual transformation on the absolute value terms in the objective function above to write $|x_i - x_j| + |y_i - y_j|$ as $(x_{ij}^+ + x_{ij}^- + y_{ij}^+ + y_{ij}^-)$, where the restrictions on these new variables may be denoted as $z \in Z$, say, where,

$$z = (x_{11}^+, \ldots, x_{n,n+m}^+, x_{11}^-, \ldots, x_{n,n+m}^-, y_{11}^+, \ldots, y_{n,n+m}^+, y_{11}^-, \ldots, y_{n,n+m}^-,$$
$$x_1, \ldots, x_n, y_1, \ldots, y_n)^t$$

and where,

$$Z = \{z: x_i - x_j - x_{ij}^+ + x_{ij}^- = 0 \quad \text{for } i=1,\ldots,n, \; j=1,\ldots,n+m$$

$$y_i - y_j - y_{ij}^+ + y_{ij}^- = 0 \quad \text{for } i=1,\ldots,n, \; j=1,\ldots,n+m$$

$$x_{ij}^+, x_{ij}^-, y_{ij}^+, y_{ij}^- \geq 0 \quad \text{for } i=1,\ldots,n, \; j=1,\ldots,n+m\}$$

Then, the rectilinear distance location-allocation problem may be written as the bilinear programming problem

$$\text{minimize} \quad c^t u + z^t T u$$

$$\text{subject to} \quad u \in U$$
$$z \in Z$$

where c and T are appropriate cost vectors and matrices respectively. Note that the orthogonality constraints of the type $x_{ij}^+ x_{ij}^- = y_{ij}^+ y_{ij}^- = 0$ are not explicitly needed since the columns of x_{ij}^+ and x_{ij}^- (as also of y_{ij}^+ and y_{ij}^-) are linearly dependent in Z.

8.3 Load Balancing Problem

The load balancing problem is one which involves the allocation of m jobs of given "weights" w_i to n departments such that the total resulting work loads, L_j, $j=1,\ldots,n$ are as equally balanced as possible. Hence, if we define 0-1 variables x_{ij} as

$$x_{ij} = \begin{cases} 1 \text{ if job 1 is assigned to department j} \\ \\ 0 \text{ otherwise} \end{cases} \qquad i=1,\ldots,m, \; j=1,\ldots,n$$

then the work load at station j is given by

$$L_j = \sum_{i=1}^{m} w_i x_{ij}$$

The concept of an equitable balance of load between stations is subjective. One may choose to minimize the difference between the minimum and the maximum work load at any station. Or, one may examine the average work load $L = \frac{1}{n} \sum_{i=1}^{m} w_i$ and choose to

$$\text{minimize} \quad \{ \sum_{j=1}^{n} |L_j - L| \}$$

Using this latter alternative one may adopt two types of formulations. Firstly, one may introduce an (m+1)th dummy job and stipulate the following constraints

$$x \epsilon X = \{x = (x_{ij}): \sum_{j=1}^{n} x_{ij} = 1, \; i=1,\ldots,m$$

$$\sum_{j=1}^{n} x_{m+1,j} = m(n-1)$$

$$\sum_{i=1}^{m+1} x_{ij} = m, \; j=1,\ldots,n$$

$$x_{ij} \geq 0\} \qquad\qquad (8.1)$$

These above constraints constitute a transportation constraint set in which there are (m+1) supply points, m of which have a unit supply and the (m+1)[st] has a

supply of m(n-1). Further, there are n demand points, each of them having a demand
of m units. Moreover, every basic feasible solution to this problem is integer,
and specifically, zero-one. In fact, there is a one-to-one correspondence between
the assignment of jobs to departments and basic feasible solutions of this constraint set. Hence, this is now an extreme point problem wherein one searches
for the best extreme point of X.

In another equivalent form, we may let

$$L_j - L = y_j^+ - y_j^-, \quad y_j^+, y_j^- \quad 0, \quad y_j^+ y_j^- = 0$$

and recalling that $L_j = \sum_{i=1}^{m} w_i x_{ij}$, we may formulate the load balancing problem as

$$\text{minimize} \quad \sum_{j=1}^{n} (y_j^+ + y_j^-)$$

$$\text{subject to} \quad \sum_{i=1}^{m} w_i x_{ij} - y_j^+ + y_j^- = L \quad j=1,\ldots,n$$

$$y_j^+, y_j^- \geq 0 \quad j=1,\ldots,n$$

and x is an extreme point of X of Equation (8.1). Note that the orthogonal constraints $y_{ij}^+ y_j^- = 0$, $j=1,\ldots,n$ may be omitted in solution procedures which set the
above problem up as linear programs since then the columns of y_j^+ and y_j^- are linearly
dependent for each $j=1,\ldots,n$.

8.4 The Segregated Storage Problem

This problem considers a certain resource which is available in quantities
$S_1,\ldots,S_m$ at m sources and is to be allocated to meet the demands $D_1,\ldots,D_n$ of n
users with the added restriction that the requirement of each of (n-1) users,
say, $1,2,\ldots,n-1$ is to be met from one and only one source. The last, or the nth,
user can be supplied from any of the sources. In a storage context, the first
(n-1) users correspond to private (special) storage facilities and the nth user

corresponds to public (common or general) storage facility. Mathematically, we may let x_{ij} denote the quantity shipped from supply point i to demand point j at a cost of say, c_{ij} per unit and formulate the problem under the assumption that $\sum_{i=1}^{m} S_i = \sum_{j=1}^{n} D_j$ as

$$\text{minimize} \quad \sum_{i=1}^{m} \sum_{j=1}^{n} c_{ij} x_{ij}$$

$$\text{subject to} \quad x \in X = \{x = (x_{ij}): \sum_{i=1}^{m} x_{ij} \le D_j, \ j=1,\ldots,n$$

$$\sum_{j=1}^{n} x_{ij} = S_i, \ i=1,\ldots,m$$

$$x_{ij} \ge 0\}$$

$$\sum_{i=1}^{m} \delta_{ij} \le 1, \ j=1,\ldots,n-1$$

where,

$$\delta_{ij} = \begin{cases} 0 & \text{if } x_{ij} = 0 \\ & \qquad \text{for } j=1,\ldots,n-1 \\ 1 & \text{if } x_{ij} > 0 \end{cases}$$

This problem is also an extreme point optimization problem since it can be shown that there exists an optimal solution to it which is an extreme point of the set X of Equation (8.2).

8.5 Production Scheduling on N-Identical Machines

Consider a firm which manufactures K products, each of which must be processed on the same machine. The machine has N rings of dies and is capable of processing N products simultaneously. Also, it is assumed that N < K. However, the entire machine must be shut down to change from one set of N products to another set. Thus, this problem may be viewed as one involving N identical machines, each with a single ring of dies, which are coupled by demand constraints. This demand is available as a forecast of each of the K products over the next T periods, with the current (initial) inventory level being known. Changeovers are permitted only at the end of each period with the cost being proportional to the number of rings whose dies must be changed. The problem is to determine an optimal production schedule for the K products over T weeks so as to minimize the total changeover and inventory costs while complying with the demand requirements.

Thus, let

x_{kt} = the number of rings producing product k in time period t,

   k=1,...,K, t=1,...,T.

d_{kt} = integral demand for product k at the end of time period t,

   k=1,...,K, t=1,...,T.

y_{kt} = inventory of product k at the end of period t, k=1,...,K,

   t=0,...,T, with t=0 yielding initial inventory.

c = cost of changing a single ring of dies.

c_k = inventory carrying cost for a single period for product k,

   k=1,...,K.

Now, it is clear that

$$\sum_{k=1}^{K} |x_{k,t+1} - x_{kt}|$$

represents twice the number of rings changed from manufacturing one product to another at the end of period t, so that the total changeover cost over T periods is

$$\frac{c}{2} \sum_{t=1}^{T-1} \sum_{k=1}^{K} |x_{k,t+1} - x_{kt}| \qquad (8.3)$$

Further, the inventory of product k during the tth period is obtaine through the cumulative occurance over t-1 periods as

$$y_{k,t-1} = y_{k0} + \sum_{i=1}^{t-1} x_{ki} - \sum_{i=1}^{t-1} d_{ki} \qquad (8.4)$$

The production constraints require all the N rings to be busy in each time period, or,

$$\sum_{k=1}^{K} x_{kt} = N, \text{ for } t=1,\ldots,T \qquad (8.5)$$

Further, to stipulate that the demands are all met with, we need the constraints

$$y_{k,t-1} + x_{kt} - y_{kt} = d_{kt} \text{ for } t=1,\ldots,T, \; k=1,\ldots,K \qquad (8.6)$$

Then, the problem at hand is to

$$\text{minimize} \quad \frac{c}{2} \sum_{t=1}^{T-1} \sum_{k=1}^{K} x_{k,t+1} - x_{kt} + \sum_{t=1}^{T} \sum_{k=1}^{K} c_k y_{k,t-1}$$

$$\text{subject to} \quad \sum_{k=1}^{K} x_{kt} = N \qquad\qquad t=1,\ldots,T$$

$$y_{k,t-1} + x_k - y_{kt} = d_{kt} \qquad t=1,\ldots,T, \; k=1,\ldots,K$$

$$x_{kt}, \; y_{kt} \geq 0, \text{ and integer, } k=1,\ldots,K, \; t=1,\ldots,T$$

To convert this problem into one with network constraints, a redundant constraint of the following form may be added

$$\sum_{k=1}^{K} y_{kT} = \text{total inventory at end of planning horizon}$$

$$= \sum_{k=1}^{K} y_{k0} + NT - \sum_{t=1}^{T} \sum_{k=1}^{K} d_{kt}$$

Utilizing the usual transformation of representing the absolute value of a variable as the difference between two nonnegative variables, the above problem may be converted into an integer linear program. The integrality restrictions may then be replaced with the equivalent requirements that the solution should be an extreme point of the network constraints given above. Hence, this problem may be represented as an extreme point optimization problem.

8.6 Fixed Charge Problem

This type of problem is a mathematical programming problem which involves a fixed cost to be added if a variable is non-zero. More specifically, the problem may be stated mathematically as

$$\text{minimize} \quad c^t x + \sum_{j=1}^{n} \delta_j f_j : x \in X$$

where X is a polyhedral set and

$$\delta_j = \begin{cases} 0 & \text{if } x_j = 0 \\ \\ 1 & \text{if } x_j > 0 \end{cases} \quad \text{for } j=1,\ldots,n$$

and where c denotes the vector of variable cost coefficients. Here f_j is the fixed charge incurred if $x_j > 0$. It can be shown that the above objective function is concave and hence there exists an optimal solution which is an extreme point of X. Thus, this too is of the class of extreme point optimization problems.

As an example of a fixed charge problem, one may think of a transportation-locatio- situation wherein the fixed cost is associated with the construction of a supply facility or a warehouse at a potential site. As another example, the fixed charge may arise as a fixed set-up cost in a scheduling problem if the decision to manufacture a certain product is adopted. Fixed charges also arise in passenger transportation models wherein the introduction of each additional transport facility involves an extra fixed cost.

8.7 Project Selection/Portfolio Allocation/Goal Programming

Consider the problem

$$\text{minimize} \quad c^t x$$

$$\text{subject to} \quad x \in S = \{x: Ax \le b_0\}$$

$$x \in \text{extreme point of } X = \{x: Bx \le b, x \ge 0\}$$

where B is block-angular with blocks $B_1,\ldots,B_p$, say.

These types of problems arise, for example, in the context of project selection problems wherein the extreme points of X correspond to projects being proposed by the p "subordinate" units. These units must be coordinated by the "superordinate" whose stipulations/restrictions are expressed by the set S. If convex combinations of projects proposed by the subordinates is not meaningful, one is restricted to selecting an extreme point of X, that is, one needs to investigate the extreme point optimization problem given above. Similar structures arise in Portfolio selection and in Goal Programming. In the latter case, the objective is to obtain a solution as "close" as possible to the preset goals.

8.8 Other Applications

In several production planning problems, one is confronted with a profit function which is convex due to economies of scale. That is, as the level of production is increased, the profits increase more rapidly than in direct proportion at first and then level off due to diminishing marginal returns. Hence the problem of maximizing a convex (often quadratic) function over linear constraints is essentially an extreme point optimization problem.

In decision theory problems, a decision tree is constructed wherein each path through the tree represents a strategy with a utility value associated with it. The objective is to maximize the expected utility over a finite set of vectors, each vector denoting the values associated with a strategy. The problem may be reduced to that of maximizing a linear function over a polytope, where the polytope is defined as the convex hull of a finite set of points. Thus this is a special case of an extreme point optimization problem where the extreme points are a subset of a known finite set of discrete points.

Finally, we note that 0-1 linear integer programming problems can be converted into problems of minimizing a concave function over a polyhedral set. This may be accomplished by simply incorporating a penalty term of the form $\sum_{j=1}^{n} M x_j(1-x_j)$ into the objective function where M is a suitably large constant, and x_j, $j=1,\ldots,n$ are variables restricted to be zero or one in value. The problem is hence an extreme point optimization problem. However, due to the ill-conditioning effects of M, usually implicit enumeration schemes have been known to permit more efficient solution procedures than the implementation of the above transformation.

8.9 Notes and References

This section elaborates on some of the problems that can be represented as disjunctive programs. The reader may note that theoretically integer-programming problems in general can be cast in a disjunctive programming format. Only some of the special practical cases, particularly with 0-1 variables, that seem more amenable to solution procedures using disjunctive programming principles are discussed in this chapter. The thought proposed and discussed by Balas [4] and Glover [19] of incorporating disjunctive programming/polyhedral annexation principles within a branch-and-bound approach is particularly significant in the context of developing viable solution procedures.

SELECTED REFERENCES

1. Balas, E., "Intersection Cuts--A New Type of Cutting Planes for Integer Programming", Operations Research, 19, 19-39, 1971.

2. Balas, E., "Integer Programming and Convex Analysis: Intersection Cuts from Outer Polars", Mathematical Programming, 2, 330-382, 1972.

3. Balas, E., "Nonconvex Quadratic Programming Via Generalized Polars", Management Science Research Report No. 278, GSIA, Carnegie-Mellon University, 1972.

4. Balas, E., "Intersection Cuts from Disjunctive Constraints", Management Science Research Report, No. 330, Carnegie-Mellon University, February 1974.

5. Balas, E., "Disjunctive Programming: Properties of the Convex Hull of Feasible Points", Management Science Research Report 348, GSIA, Carnegie-Mellon University, 1974.

6. Balas, E., "Disjunctive Programming: Cutting Planes from Logical Conditions", in Nonlinear Programming, O. L. Mangasarian, R. R. Meyer, and S. M. Robinson, eds., Academic Press, New York, 1975.

7. Balas, E., "Disjunctive Programming", Management Science Research Report No. 415, Carnegie-Mellon University, August 1977.

8. Balas, E., and C. A. Burdet, "Maximizing a Convex Quadratic Function Subject to Linear Constraints", Management Science Research Report No. 299, GSIA, Carnegie-Mellon University, 1973.

9. Bazaraa, M. S. and C. M. Shetty, "Nonlinear Programming: Theory and Algorithms", John Wiley and Sons, New York, 1979.

10. Burdet, C. A., "Polaroids: A New Tool in Nonconvex and in Integer Programming", Naval Research Logistics Quarterly, 20, 13-22, 1973.

11. Burdet, C. A., "On Polaroid Intersections", in Mathematical Programming in Theory and Practice, P. Hammer and G. Zoutendijk, eds., North Holland, 1974.

12. Burdet, C. A., "Elements of a Theory in Nonconvex Programming", Naval Research Logistics Quarterly, Volume 24, No. 1, pp. 47-66, 1977.

13. Burdet, C. A., "Convex and Polaroid Extensions", Naval Research Logistics Quarterly, Volume 26, No. 1, pp. 67-82, 1977.

14. Dem'janov, V. F., "Seeeking a Minimax on a Bounded Set", Soviet Mathematics Doklady, Volume 11, pp. 517-521, 1970 (English Translation).

15. Glover, F. and D. Klingman, "The Generalized Lattice Point Problem", Management Science Research Report 71-3, University of Colorado, 1971.

16. Glover, F., "Convexity Cuts and Cut Search", Operations Research, 21, 123-134, 1973.

17. Glover, F., "Convexity Cuts for Multiple Choice Problems", Discrete Mathematics, Volume 6, pp. 221-234, 1973.

18. Glover, F., "Polyhedral Convexity Cuts and Negative Edge Extensions", Zeitschrift für Operations Research, 18, pp. 181-186, 1974.

19. Glover, F., "Polyhedral Annexation in Mixed Integer and Combinatorial Programming", _Mathematical Programming_, Vol. 8, pp. 161-188, 1975. (See also MSRS Report 73-9, University of Colorado, August 1973).

20. Glover, F., D. Klingman and J. Stutz, "The Disjunctive Facet Problem: Formulation and Solution Techniques", _Management Science Research Report_, No. 72-10, University of Colorado, 1972.

21. Grünbaum, B., "Convex Polytopes", Interscience, New York, 1967.

22. Held, M., P. Wolfe, and H. D. Crowder, "Validation of Subgradient Optimization", _Mathematical Programming_, Volume 6, pp. 62-88, 1974.

23. Jeroslow, R. G., "The Principles of Cutting Plane Theory: Part I", (with an addendum), GSIA, Carnegie-Mellon University, February, 1974.

24. Jeroslow, R. G., "Cutting Planes for Complementarity Constraints", _SIAM Journal of Control and Optimization_, Volume 16, No. 1, pp. 56-62, 1976.

25. Jeroslow, R. G., "Cutting Plane Theory: Disjunctive Methods", _Annals of Discrete Mathematics_, Volume 1, pp. 293-330, 1977.

26. Jeroslow, R. G., "A Cutting Plane Game and its Algorithms", Discussion Paper No. 7724, _Center for Operations Research and Econometrics_, University of Catholique de Louvain, June 1977.

27. Karlin, S., "Mathematical Methods and Theory in Games, Programming and Economics", Volume 1, Addison-Wesley Publishing Company, Reading, Mass., 1959.

28. Majthay, A. and A. Whinston, "Quasi-Concave Minimization Subject to Linear Constraints", _Discrete Mathematics_, Volume 9, pp. 35-59, 1974.

29. Owen, G., "Cutting Planes for Programs with Disjunctive Constraints", _Optimization Theory and Its Applications_, Volume 11, pp. 49-55, 1973.

30. Poljak, B. T., "A General Method of Solving Extremum Problems", _Soviet Mathematics Doklady_, Volume 8, pp. 593-597, 1967. (English Translation).

31. Poljak, B. T., "Minimization of Unsmooth Functionals", _USSR Computational Mathematics and Mathematical Physics_, Volume 9, pp. 14-29, 1969. (English Translation).

32. Sherali, H. D. and C. M. Shetty, "Deep Cuts in Disjunctive Programming", _Naval Research Logistics Quarterly_, Volume 27, pp. 453-475, 1980.

33. Sherali, H. D. and C. M. Shetty, "A Finitely Convergent Algorithm for Bilinear Programming Problems Using Polar Cuts and Disjunctive Face Cuts", _Math Programming_, to appear.

34. Sherali, H. D. and C. M. Shetty, "Disjunctive Programming, Polyhedral Annexation Techniques, and Nondominated Disjunctive Cutting Planes", Georgia Institute of Technology, 1979.

35. Sherali, H. D. and C. M. Shetty, "A Finitely Convergent Procedure for Facial Disjunctive Programs", Georgia Institute of Technology, 1979.

36. Vaish, H. and C. M. Shetty, "A Cutting Plane Algorithm for the Bilinear Programming Problem", _Naval Research Logistics Quarterly_, Volume 24, No. 1, pp. 83-94, March 1975.

Vol. 83: NTG/GI-Gesellschaft für Informatik, Nachrichtentechnische Gesellschaft. Fachtagung „Cognitive Verfahren und Systeme", Hamburg, 11.–13. April 1973. Herausgegeben im Auftrag der NTG/GI von Th. Einsele, W. Giloi und H.-H. Nagel. VIII, 373 Seiten. 1973.

Vol. 84: A. V. Balakrishnan, Stochastic Differential Systems I. Filtering and Control. A Function Space Approach. V, 252 pages. 1973.

Vol. 85: T. Page, Economics of Involuntary Transfers: A Unified Approach to Pollution and Congestion Externalities. XI, 159 pages. 1973.

Vol. 86: Symposium on the Theory of Scheduling and its Applications. Edited by S. E. Elmaghraby. VIII, 437 pages. 1973.

Vol. 87: G. F. Newell, Approximate Stochastic Behavior of n-Server Service Systems with Large n. VII, 118 pages. 1973.

Vol. 88: H. Steckhan, Güterströme in Netzen. VII, 134 Seiten. 1973.

Vol. 89: J. P. Wallace and A. Sherret, Estimation of Product. Attributes and Their Importances. V, 94 pages. 1973.

Vol. 90: J.-F. Richard, Posterior and Predictive Densities for Simultaneous Equation Models. VI, 226 pages. 1973.

Vol. 91: Th. Marschak and R. Selten, General Equilibrium with Price-Making Firms. XI, 246 pages. 1974.

Vol. 92: E. Dierker, Topological Methods in Walrasian Economics. IV, 130 pages. 1974.

Vol. 93: 4th IFAC/IFIP International Conference on Digital Computer Applications to Process Control, Part I. Zürich/Switzerland, March 19–22, 1974. Edited by M. Mansour and W. Schaufelberger. XVIII, 544 pages. 1974.

Vol. 94: 4th IFAC/IFIP International Conference on Digital Computer Applications to Process Control, Part II. Zürich/Switzerland, March 19–22, 1974. Edited by M. Mansour and W. Schaufelberger. XVIII, 546 pages. 1974.

Vol. 95: M. Zeleny, Linear Multiobjective Programming. X, 220 pages. 1974.

Vol. 96: O. Moeschlin, Zur Theorie von Neumannscher Wachstumsmodelle. XI, 115 Seiten. 1974.

Vol. 97: G. Schmidt, Über die Stabilität des einfachen Bedienungskanals. VII, 147 Seiten. 1974.

Vol. 98: Mathematical Methods in Queueing Theory. Proceedings 1973. Edited by A. B. Clarke. VII, 374 pages. 1974.

Vol. 99: Production Theory. Edited by W. Eichhorn, R. Henn, O. Opitz, and R. W. Shephard. VIII, 386 pages. 1974.

Vol. 100: B. S. Duran and P. L. Odell, Cluster Analysis. A Survey. VI, 137 pages. 1974.

Vol. 101: W. M. Wonham, Linear Multivariable Control. A Geometric Approach. X, 344 pages. 1974.

Vol. 102: Analyse Convexe et Ses Applications. Comptes Rendus, Janvier 1974. Edited by J.-P. Aubin. IV, 244 pages. 1974.

Vol. 103: D. E. Boyce, A. Farhi, R. Weischedel, Optimal Subset Selection. Multiple Regression, Interdependence and Optimal Network Algorithms. XIII, 187 pages. 1974.

Vol. 104: S. Fujino, A Neo-Keynesian Theory of Inflation and Economic Growth. V, 96 pages. 1974.

Vol. 105: Optimal Control Theory and its Applications. Part I. Proceedings 1973. Edited by B. J. Kirby. VI, 425 pages. 1974.

Vol. 106: Optimal Control Theory and its Applications. Part II. Proceedings 1973. Edited by B. J. Kirby. VI, 403 pages. 1974.

Vol. 107: Control Theory, Numerical Methods and Computer Systems Modeling. International Symposium, Rocquencourt, June 17–21, 1974. Edited by A. Bensoussan and J. L. Lions. VIII, 757 pages. 1975.

Vol. 108: F. Bauer et al., Supercritical Wing Sections II. A Handbook. V, 296 pages. 1975.

Vol. 109: R. von Randow, Introduction to the Theory of Matroids. IX, 102 pages. 1975.

Vol. 110: C. Striebel, Optimal Control of Discrete Time Stochastic Systems. III. 208 pages. 1975.

Vol. 111: Variable Structure Systems with Application to Economics and Biology. Proceedings 1974. Edited by A. Ruberti and R. R. Mohler. VI, 321 pages. 1975.

Vol. 112: J. Wilhelm, Objectives and Multi-Objective Decision Making Under Uncertainty. IV, 111 pages. 1975.

Vol. 113: G. A. Aschinger, Stabilitätsaussagen über Klassen von Matrizen mit verschwindenden Zeilensummen. V, 102 Seiten. 1975.

Vol. 114: G. Uebe, Produktionstheorie. XVII, 30? Seiten. 1976.

Vol. 115: Anderson et al., Foundations of System Theory: Finitary and Infinitary Conditions. VII, 93 pages. 1976

Vol. 116: K. Miyazawa, Input-Output Analysis and the Structure of Income Distribution. IX, 135 pages. 1976.

Vol. 117: Optimization and Operations Research. Proceedings 1975. Edited by W. Oettli and K. Ritter. IV, 316 pages. 1976.

Vol. 118: Traffic Equilibrium Methods, Proceedings 1974. Edited by M. A. Florian. XXIII, 432 pages. 1976.

Vol. 119: Inflation in Small Countries. Proceedings 1974. Edited by H. Frisch. VI, 356 pages. 1976.

Vol. 120: G. Hasenkamp, Specification and Estimation of Multiple-Output Production Functions. VII, 151 pages. 1976.

Vol. 121: J. W. Cohen, On Regenerative Processes in Queueing Theory. IX, 93 pages. 1976.

Vol. 122: M. S. Bazaraa, and C. M. Shetty, Foundations of Optimization VI. 193 pages. 1976

Vol. 123: Multiple Criteria Decision Making. Kyoto 1975. Edited by M. Zeleny. XXVII, 345 pages. 1976.

Vol. 124: M. J. Todd. The Computation of Fixed Points and Applications. VII, 129 pages. 1976.

Vol. 125: Karl C. Mosler. Optimale Transportnetze. Zur Bestimmung ihres kostengünstigsten Standorts bei gegebener Nachfrage. VI, 142 Seiten. 1976.

Vol. 126: Energy, Regional Science and Public Policy. Energy and Environment I. Proceedings 1975. Edited by M. Chatterji and P. Van Rompuy. VIII, 316 pages. 1976.

Vol. 127: Environment, Regional Science and Interregional Modeling. Energy and Environment II. Proceedings 1975. Edited by M. Chatterji and P. Van Rompuy. IX, 211 pages. 1976.

Vol. 128: Integer Programming and Related Areas. A Classified Bibliography. Edited by C. Kastning. XII, 495 pages. 1976.

Vol. 129: H.-J. Lüthi, Komplementaritäts- und Fixpunktalgorithmen in der mathematischen Programmierung. Spieltheorie und Ökonomie. VII, 145 Seiten. 1976.

Vol. 130: Multiple Criteria Decision Making, Jouy-en-Josas, France. Proceedings 1975. Edited by H. Thiriez and S. Zionts. VI, 409 pages. 1976.

Vol. 131: Mathematical Systems Theory. Proceedings 1975. Edited by G. Marchesini and S. K. Mitter. X, 408 pages. 1976.

Vol. 132: U. H. Funke, Mathematical Models in Marketing. A Collection of Abstracts. XX, 514 pages. 1976.

Vol. 133: Warsaw Fall Seminars in Mathematical Economics 1975. Edited by M. W. Loś, J. Loś, and A. Wieczorek. V, 159 pages. 1976.

Vol. 134: Computing Methods in Applied Sciences and Engineering. Proceedings 1975. VIII, 390 pages. 1976.

Vol. 135: H. Haga, A Disequilibrium – Equilibrium Model with Money and Bonds. A Keynesian – Walrasian Synthesis. VI, 119 pages. 1976.

Vol. 136: E. Kofler und G. Menges, Entscheidungen bei unvollständiger Information. XII, 357 Seiten. 1976.

Vol. 137: R. Wets, Grundlagen Konvexer Optimierung. VI, 146 Seiten. 1976.

Vol. 138: K. Okuguchi, Expectations and Stability in Oligopoly Models. VI, 103 pages. 1976.

Vol. 139: Production Theory and Its Applications. Proceedings. Edited by H. Albach and G. Bergendahl. VIII, 193 pages. 1977.

Vol. 140: W. Eichhorn and J. Voeller, Theory of the Price Index. Fisher's Test Approach and Generalizations. VII, 95 pages. 1976.

Vol. 141: Mathematical Economics and Game Theory. Essays in Honor of Oskar Morgenstern. Edited by R. Henn and O. Moeschlin. XIV, 703 pages. 1977.

Vol. 142: J. S. Lane, On Optimal Population Paths. V, 123 pages. 1977.

Vol. 143: B. Näslund, An Analysis of Economic Size Distributions. XV, 100 pages. 1977.

Vol. 144: Convex Analysis and Its Applications. Proceedings 1976. Edited by A. Auslender. VI, 219 pages. 1977.

Vol. 145: J. Rosenmüller, Extreme Games and Their Solutions. IV, 126 pages. 1977.

Vol. 146: In Search of Economic Indicators. Edited by W. H. Strigel. XVI, 198 pages. 1977.

Vol. 147: Resource Allocation and Division of Space. Proceedings. Edited by T. Fujii and R. Sato. VIII, 184 pages. 1977.

Vol. 148: C. E. Mandl, Simulationstechnik und Simulationsmodelle in den Sozial- und Wirtschaftswissenschaften. IX, 173 Seiten. 1977.

Vol. 149: Stationäre und schrumpfende Bevölkerungen: Demographisches Null- und Negativwachstum in Österreich. Herausgegeben von G. Feichtinger. VI, 262 Seiten. 1977.

Vol. 150: Bauer et al., Supercritical Wing Sections III. VI, 179 pages. 1977.

Vol. 151: C. A. Schneeweiß, Inventory-Production Theory. VI, 116 pages. 1977.

Vol. 152: Kirsch et al., Notwendige Optimalitätsbedingungen und ihre Anwendung. VI, 157 Seiten. 1978.

Vol. 153: Kombinatorische Entscheidungsprobleme: Methoden und Anwendungen. Herausgegeben von T. M. Liebling und M. Rössler. VIII, 206 Seiten. 1978.

Vol. 154: Problems and Instruments of Business Cycle Analysis. Proceedings 1977. Edited by W. H. Strigel. VI, 442 pages. 1978.

Vol. 155: Multiple Criteria Problem Solving. Proceedings 1977. Edited by S. Zionts. VIII, 567 pages. 1978.

Vol. 156: B. Näslund and B. Sellstedt, Neo-Ricardian Theory. With Applications to Some Current Economic Problems. VI, 165 pages. 1978.

Vol. 157: Optimization and Operations Research. Proceedings 1977. Edited by R. Henn, B. Korte, and W. Oettli. VI, 270 pages. 1978.

Vol. 158: L. J. Cherene, Set Valued Dynamical Systems and Economic Flow. VIII, 83 pages. 1978.

Vol. 159: Some Aspects of the Foundations of General Equilibrium Theory: The Posthumous Papers of Peter J. Kalman. Edited by J. Green. VI, 167 pages. 1978.

Vol. 160: Integer Programming and Related Areas. A Classified Bibliography. Edited by D. Hausmann. XIV, 314 pages. 1978.

Vol. 161: M. J. Beckmann, Rank in Organizations. VIII, 164 pages. 1978.

Vol. 162: Recent Developments in Variable Structure Systems, Economics and Biology. Proceedings 1977. Edited by R. R. Mohler and A. Ruberti. VI, 326 pages. 1978.

Vol. 163: G. Fandel, Optimale Entscheidungen in Organisationen. VI, 143 Seiten. 1979.

Vol. 164: C. L. Hwang and A. S. M. Masud, Multiple Objective Decision Making – Methods and Applications. A State-of-the-Art Survey. XII, 351 pages. 1979.

Vol. 165: A. Maravall, Identification in Dynamic Shock-Error Models. VIII, 158 pages. 1979.

Vol. 166: R. Cuninghame-Green, Minimax Algebra. XI, 258 pages. 1979.

Vol. 167: M. Faber, Introduction to Modern Austrian Capital Theory. X, 196 pages. 1979.

Vol. 168: Convex Analysis and Mathematical Economics. Proceedings 1978. Edited by J. Kriens. V, 136 pages. 1979.

Vol. 169: A. Rapoport et al., Coalition Formation by Sophisticated Players. VII, 170 pages. 1979.

Vol. 170: A. E. Roth, Axiomatic Models of Bargaining. V, 121 pages. 1979.

Vol. 171: G. F. Newell, Approximate Behavior of Tandem Queues. XI, 410 pages. 1979.

Vol. 172: K. Neumann and U. Steinhardt, GERT Networks and the Time-Oriented Evaluation of Projects. 268 pages. 1979.

Vol. 173: S. Erlander, Optimal Spatial Interaction and the Gravity Model. VII, 107 pages. 1980.

Vol. 174: Extremal Methods and Systems Analysis. Edited by A. V. Fiacco and K. O. Kortanek. XI, 545 pages. 1980.

Vol. 175: S. K. Srinivasan and R. Subramanian, Probabilistic Analysis of Redundant Systems. VII, 356 pages. 1980.

Vol. 176: R. Färe, Laws of Diminishing Returns. VIII, 97 pages. 1980.

Vol. 177: Multiple Criteria Decision Making-Theory and Application. Proceedings, 1979. Edited by G. Fandel and T. Gal. XVI, 570 pages. 1980.

Vol. 178: M. N. Bhattacharyya, Comparison of Box-Jenkins and Bonn Monetary Model Prediction Performance. VII, 146 pages. 1980.

Vol. 179: Recent Results in Stochastic Programming. Proceedings, 1979. Edited by P. Kall and A. Prékopa. IX, 237 pages. 1980.

Vol. 180: J. F. Brotchie, J. W. Dickey and R. Sharpe, TOPAZ – General Planning Technique and its Applications at the Regional, Urban, and Facility Planning Levels. VII, 356 pages. 1980.

Vol. 181: H. D. Sherali and C. M. Shetty, Optimization with Disjunctive Constraints. VIII, 156 pages. 1980.

Ökonometrie und Unternehmensforschung
Econometrics and Operations Research

Vol. I   Nichtlineare Programmierung. Von H. P. Künzi und W. Krelle unter Mitwirkung von W. Oettli. Vergriffen

Vol. II   Lineare Programmierung und Erweiterungen. Von G. B. Dantzig. Ins Deutsche übertragen und bearbeitet von A. Jaeger. – Mit 103 Abbildungen. XVI, 712 Seiten. 1966. Geb.

Vol. III Stochastic Processes. By M. Girault. – With 35 figures. XII, 126 pages. 1966. Cloth

Vol. IV   Methoden der Unternehmensforschung im Versicherungswesen. Von K. H. Wolff. – Mit 14 Diagrammen. VIII, 266 Seiten. 1966. Geb.

Vol. V   The Theory of Max-Min and its Application to Weapons Allocation Problems. By John M. Danskin. – With 6 figures. X, 126 pages. 1967. Cloth

Vol. VI   Entscheidungskriterien bei Risiko. Von H. Schneeweiss. – Mit 35 Abbildungen. XII, 215 Seiten. 1967. Geb.

Vol. VII Boolean Methods in Operations Research and Related Areas. By P. L. Hammer (Ivănescu) and S. Rudeanu. With a preface by R. Bellman. – With 25 figures. XVI, 329 pages. 1968. Cloth

Vol. VIII Strategy for R & D: Studies in the Microeconomics of Development. By Th. Marschak. Th K. Glennan Jr., and R. Summers. – With 44 figures. XIV, 330 pages. 1967. Cloth

Vol. IX   Dynamic Programming of Economic Decisions. By M. J. Beckmann. – With 9 figures XII, 143 pages. 1968. Cloth

Vol. X   Input-Output-Analyse. Von J. Schumann. – Mit 12 Abbildungen. X, 311 Seiten. 1968. Geb.

Vol. XI   Produktionstheorie. Von W. Wittmann. – Mit 54 Abbildungen. VIII, 177 Seiten. 1968. Geb.

Vol. XII Sensitivitätsanalysen und parametrische Programmierung. Von W. Dinkelbach. – Mit 20 Abbildungen. XI, 190 Seiten. 1969. Geb.

Vol. XIII Graphentheoretische Methoden und ihre Anwendungen. Von W. Knödel. – Mit 24 Abbildungen. VIII, 111 Seiten. 1969. Geb.

Vol. XIV Praktische Studien zur Unternehmensforschung. Von E. Nievergelt, O. Müller, F. E. Schlaepfer und W. H. Landis. – Mit 82 Abbildungen. XII, 240 Seiten. 1970. Geb.

Vol. XV   Optimale Reihenfolgen. Von H. Müller-Merbach. – Mit 45 Abbildungen. IX, 225 Seiten. 1970. Geb.

Vol. XVI Preispolitik der Mehrproduktenunternehmung in der statischen Theorie. Von R. Selten. – Mit 20 Abbildungen. VIII, 195 Seiten. 1970. Geb.

Vol. XVII Information Theory for Systems Engineers. By L. P. Hyvärinen. – With 42 figures. VIII, 197 pages. 1970. Cloth

Vol. XVIII Unternehmensforschung im Bergbau. Von F. L. Wilke. – Mit 29 Abbildungen. VIII, 150 Seiten. 1972. Geb.

Vol. XIX Anti-Äquilibrium. Von J. Kornai. – Mit 31 Abbildungen. XVI, 382 Seiten. 1975. Geb.

Vol. XX   Mathematische Optimierung. Von E. Blum, W. Oettli. IX, 413 Seiten. (davon 75 Seiten Bibliographie). 1975. Geb.

Vol. XXI Stochastic Linear Programming. By P. Kall. VI, 95 pages. 1976. Cloth.